# Lecture Notes in Mathematics

A collection of informal reports and seminars
Edited by A. Dold, Heidelberg and B. Eckmann, Zürich



58

# Kuramochi Boundaries of Riemann Surfaces

A Symposium held at the Research Institute for Mathematical Sciences, Kyoto University, October 1965

1968

Springer-Verlag Berlin · Heidelberg · New York

Library of Congress Catalog Card Number 68-26746. Printed in Germany. Title No. 3664

# F O R E W O R D

The papers included in this monograph were presented at a symposium held at the Research Institute for Mathematical Sciences of Kyoto University, October 11-13, 1965. The proceedings were edited by F-Y. Maeda and M. Ohtsuka and several of the papers were revised after the symposium.

We owe this publication to a suggestion by Professor Doob. Appreciations are due also to Professor Kusunoki for his coordination of the symposium and to the Institute for its financial support.

Hiroshima

January 25, 1968

Fumi-Yuki Maeda

Makoto Ohtsuka

C O N T E N T S

# I. INTRODUCTION TO THE KURAMOCHI BOUNDARY

Fumi-Yuki MAEDA

## Introduction

In [3], Z. Kuramochi introduced an ideal boundary of an open Riemann surface having nice function-theoretic properties. While Kuramochi himself continued to develop his theory, C. Constantinescu and A. Cornea picked up his idea and reformed the theory in very systematical form ([1]). Their work clarified the importance of this boundary, which is now called the Kuramochi boundary.

In their book [1], Constantinescu and Cornea remarked that the Kuramochi boundary points share many properties enjoyed by interior points. This rather obscure remark may be interpreted in various ways. One way is to observe that, for the Kuramochi boundary, we can consider a kernel like Green function whose pole lies on the boundary and we can develop a potential theory on the space including the boundary.

Now suppose that we want an ideal boundary which possesses the properties mentioned in the above paragraph, without knowing the Kuramochi boundary and let us consider the problem of constructing such an ideal boundary. We shall denote the base space by R. It may be any open Riemann surface or any non-compact space of type $\mathcal{E}$ in the sense of Brelot-Choquet. In considering an ideal boundary $\Delta$ of R, we here restrict ourselves to the case where $\Delta \cup R$ is compact, i.e., it is a compactification of R. Thus, we first make a review on general methods of compactification.

## § 1. The Q-compactification

In this section, let X be a locally compact Hausdorff space. A compactification $\hat{X}$ of X is a compact Hausdorff space which contains X as a dense open subset. $\Delta = \hat{X} - X$ is called an ideal boundary of X.

Let Q be a family of bounded continuous functions on X. Then there exists a unique (up to a homeomorphism) compactification $\hat{X}$ satisfying the following two conditions:

a) Each $f \in Q$ can be continuously extended over $\hat{X}$;

b) Q separates points of $\Delta = \hat{X} - X$, i.e., for any $x, y \in \Delta$, $x \neq y$, there exists $f \in Q$ such that $f(x) \neq f(y)$.

This compactification is called the *Q-compactification* of X and will be denoted by $\hat{X}_Q$ ([1], § 9). If $C_0$ is the space of all continuous functions with compact support on X, then it is obvious that $\hat{X}_{Q \cup C_0} = \hat{X}_Q$. There are various methods to realize $\hat{X}_Q$, if Q is of special type. For example, if Q is an algebra containing constants and $C_0$, then $\hat{X}_Q$ coincides with the space of characters on Q or the maximal ideal space of Q. Also, the following point of view is useful.

Let Q contain $C_0$. For any finite subfamily F of Q and for any $\varepsilon > 0$, let

$$V_{F,\varepsilon} = \{(x, y) \in X \times X; \ |f(x) - f(y)| < \varepsilon \ \text{ for all } f \in F\}.$$

The collection $\mathcal{V}_Q$ of all such sets $V_{F,\varepsilon}$ defines a uniform structure on X, which is compatible with the original topology in X. (See [2] for general informations on uniform structures.) Then it is easy to see that the completion of $(X, \mathcal{V}_Q)$ coincides with $\hat{X}_Q$. (Hence $(X, \mathcal{V}_Q)$ is precompact and $\mathcal{V}_Q$ is the uniform structure compatible with the topology of $\hat{X}_Q$.)

If Q is separable with respect to the sup-norm, i.e., if we can find a sequence $\{f_i\} \subseteq Q$ which is dense in Q, then we define a distance d on X by

$$d(x, y) = \sum_{i=1}^{\infty} \frac{1}{2^i} \frac{|f_i(x) - f_i(y)|}{1 + |f_i(x) - f_i(y)|}$$

The uniform structure induced by d is nothing but the structure $\mathcal{V}_Q$. Hence the completion of (X, d) can be regarded as the Q-compactification $\hat{X}_Q$, so that $\hat{X}_Q$ is metrizable in this case.

## § 2. Definition by Constantinescu and Cornea

We have seen that a particular ideal boundary of R can be realized by choosing a suitable family Q of bounded continuous functions on R. Thus our next question is to find Q such that $\hat{R}_Q$ satisfies the properties stated in the introduction. To cast a light on this question, let us consider a special case where R is the interior of a bordered Riemann surface. In this case the required ideal boundary should coincide with the border B. Hence, functions in Q must be continuous when it is symmetrically extended over the double of R and also Q must separate points on B. Thus we seek for a method of defining such a family Q which can be applied for a general R. In function theory or in potential theory, we are ready to try with harmonic functions. If we consider the family of functions which are harmonic near the border when symmetrically extended over the double of R, then this family determines the border. Such a function f can be characterized as follows: Let K be a compact set in R outside of which f is harmonic. Then f has the smallest Dirichlet integral on R - K among the functions assuming the values f on K. This last formulation can be given for any (not necessarily bordered) space. Precisely, we must use the following form of Dirichlet Principle:

*Dirichlet Principle.* Let K be a non-polar compact set in R and let f be a continuously differentiable function defined in a neighborhood of K. Then there exists a unique function $f^K$ on R - K satisfying the following properties:

(i) $f^K$ is harmonic on R - K;

(ii) $D_{R-K}(f^K) = \inf \{D_{R-K}(g);\ D_R(g) < \infty,\ f = g \text{ on } K\}$;

(iii) $\lim_{p \to q,\, p \in R-K} f^K(p) = f(q)$ for every regular boundary point q of R - K.

Here $D_A(g)$ denotes the Dirichlet integral of a continuously differentiable function g on a Borel set A.

When R is a Riemann surface, this theorem is a corollary to Satz 15.1 of [1]. The same method of the proof can be applied to the case when R is a space of type $\mathcal{E}$ (cf. [7]). A more or less direct proof of this principle (with a little more restrictive form) is given in [8].

It is clear that functions $f^K$ have the required property as members of Q. Thus we consider the family $\mathfrak{N}$ of all bounded countinuous functions f on R for each of which there exists a non-polar compact set K such that $f = f^K$ on R - K. We shall show that the $\mathfrak{N}$-compactification $\hat{R}_{\mathfrak{N}}$ actually satisfies the properties stated in the introduction, partly by proving that $\Delta_{\mathfrak{N}} = \hat{R}_{\mathfrak{N}} - R$ coincides with the ideal boundary originally given by Kuramochi. Constantinescu and Cornea called $\hat{R}_{\mathfrak{N}}$ the *Kuramochi compactification of* R and $\Delta_{\mathfrak{N}}$ the *Kuramochi boundary*. ([1]; also see [7].)

## § 3. Kernel N(z, p)

Our next task is to find a kernel $N_p(z) = N(z, p)$ which can be naturally defined also for $p \in \Delta_{\mathfrak{N}}$ in such a way that $N_p$ behaves like a Green function. As is easily seen from the special case of bordered Riemann surface, it is necessary to remove a hole $K_0$ from R to consider such a kernel. More precisely, let $R \cup B$ be a bordered Riemann surface, let $K_0$ be a closed disk in R and let $R' = R - K_0$. The double $\tilde{R}'$ of R' with respect to B has the Green function $\tilde{G}(z, p)$, which vanishes on

$\partial K_0$ and on its symmetric part. In order to obtain a kernel which is symmetric on $\tilde{R}'$ and is like $\tilde{G}(z, p)$ on $R'$, it is enough to consider $\tilde{G}(z, p) + \tilde{G}(z, p^*)$ for $p \in R'$, where $p^*$ is the symmetric point of p. If $p \in B$, then it is equal to $2\tilde{G}(z, p)$. Thus, in this case, we may define $N(z, p) = \tilde{G}(z, p) + \tilde{G}(z, p^*)$. Incidentally, the Green function $G'(z, p)$ of $R'$ is given by $G'(z, p) = \tilde{G}(z, p) - \tilde{G}(z, p^*)$. Hence $N(z, p) - G'(z, p) = 2\tilde{G}(z, p^*) \geqq 0$.

The kernel N can be characterized by the following properties, which do not refer to the double or to the border of R:

a) $N(z, p) - G'(z, p)$ is harmonic in $z \in R'$ for each $p \in R'$;

b) $N(z, p) = N(p, z)$;

c) $\lim_{z \to \partial K_0} N_p(z) = 0$ for each $p \in R'$;

d) If K is a compact set in R containing $K_0 \cup \{p\}$ in its interior, then $N_p{}^K = N_p$ on $R - K$.

Now, let R be arbitrary, let $K_0$ be a closed sphere in R and let $R' = R - K_0$. Then we can show the existence and uniqueness of a function $N(z, p)$ satisfying the properties a) - - d). For a Riemann surface, Kuramochi constructed it using an exhaustion of R. ([3] and [4]. Also see [8] for more rigorous treatments.) Constantinescu and Cornea ([1], § 15) showed the existence, by constructing a kernel which corresponds to $2\tilde{G}(z, p^*)$ in the case of a bordered Riemann surface. The latter construction, which uses the theory of Dirichlet functions (BLD-functions), can be readily extended to the case where R is a space of type $\mathcal{E}$. In fact, let $HD_0 = \{u;$ harmonic on $R'$, $D_{R'}(u) < \infty$ and $\lim_{p \to \partial K_0} u(p) = 0\}$. Then $HD_0$ becomes a Hilbert space with respect to the mutual Dirichlet integral $\langle u, v\rangle$. Hence, for each $p \in R'$, there exists $U_p \in HD_0$ such that $\langle U_p, u\rangle = \omega u(p)$, $\omega$ being a constant depending only on the dimension of R. We can show that $U(z, p) = U_p(z)$ corresponds to $2\tilde{G}(z, p^*)$ and $N(z, p) = G'(z, p) + U(z, p)$ is the required kernel. (See [1], § 15 for details.) By the properties b) and d) of N, N(z, p) can be naturally extended for $p \in \Delta_{\mathfrak{N}}$ and in fact it is defined to be continuous on $R' \times \hat{R}'_{\mathfrak{N}}$ ($\hat{R}'_{\mathfrak{N}} = R' \cup \Delta_{\mathfrak{N}} = \hat{R}_{\mathfrak{N}} - K_o$).

## § 4. Equivalence to Kuramochi's definition

Let us recall Kuramochi's definition of his ideal boundary [3, 4 and 6]. A sequence $\{p_i\}$ of points in R' is called fundamental if it has no limit point in R and $\{N_{p_i}\}$ converges to a harmonic function on R'. Two fundamental sequences $\{p_i\}$ and $\{p'_i\}$ are said to be equivalent if $\lim_i N_{p_i} = \lim_i N_{p'_i}$. The equivalence classes of fundamental sequences are defined to form the ideal boundary Δ and the topology of $\bar{R}' = R' \cup \Delta$ is given by the distance

$$\delta(p_1, p_2) = \sup_{z \in R_1} \left| \frac{N_{p_1}(z)}{1 + N_{p_1}(z)} - \frac{N_{p_2}(z)}{1 + N_{p_2}(z)} \right| ,$$

where $R_1$ is a relatively compact domain of R containing $K_o$ and, for $q \in \Delta$, $N_q = \lim N_{p_i}$ with a fundamental sequence $\{p_i\}$ in the class q. If we take a metric $\delta_1$ on R which is compatible with the original topology of R and is equivalent to δ on R', then $\{p_i\}$ ($p_i \in R'$) is a fundamental sequence if and only if it is a Cauchy sequence with respect to $\delta_1$ and it has no limit point in R. Hence $\bar{R} = R \cup \Delta$ is nothing but the completion of $(R, \delta_1)$.

Now, let

$$f_z(p) = \begin{cases} \dfrac{N_p(z)}{1 + N_p(z)} & \text{if } p \in R' \\ 0 & \text{if } p \in K_o \end{cases}$$

for $z \in R'$. Then each $f_z$ is a bounded continuous function on R. Let G be any open set in R' and let $\mathfrak{N}(G) = \{f_z;\ z \in G\}$, $\mathfrak{N}_1(G) = \mathfrak{N}(G) \cup C_o$. Then $\mathfrak{N}_1(R_1)$-uniform structure $\mathcal{U}_{\mathfrak{N}_1(R_1)}$ coincides with $\delta_1$-uniform structure (cf. § 1). Therefore $\bar{R}$ is the completion of $(R, \wp_{\mathcal{N}_1}(R_1))$. On the other hand, the harmonicity of functions $N_p$ on $R' - \{p\}$ implies that $\wp_{\mathcal{N}_1}(R_1)$ coincides with $\wp_{\mathcal{N}_1}(G)$ for any G, in particular, with $\wp_{\mathcal{N}_1}(R')$. Hence, from the arguments in § 1, we see that

$$\begin{aligned}\bar{R} &= \text{the completion of } (R, \wp_{\mathcal{N}_1}(R')) \\ &= \hat{R}_{\mathcal{N}_1}(R') = \hat{R}_{\mathcal{N}}(R').\end{aligned}$$

Thus, to show that $\bar{R} = \hat{R}_{\mathcal{N}}$, i.e., that Constantinescu-Cornea's definition coincides with Kuramochi's, it is enough to prove that $\hat{R}_{\mathcal{N}} = \hat{R}_{\mathcal{N}}(R')$. We give here a sketch of its proof which is due to Constantinescu and Cornea ([1], § 16).

Since $f_z$ $(z \in R')$ can be continuously extended over $\hat{R}_{\mathcal{N}}$, it is enough to show that, for any $q_1, q_2 \in \Delta_{\mathcal{N}}$ $(q_1 \neq q_2)$, there exists $z \in R'$ such that $N(z, q_1) \neq N(z, q_2)$. By definition, there exists $g \in \mathcal{N}$ such that $g(q_1) \neq g(q_2)$. Since g is harmonic outside a compact set K, we can find $g_1 \in C^{\infty}(R)$ such that $g_1 = 0$ on $K_o$, $g_1 = g$ outside a compact set $K' \supseteq K \cup K_o$ and $g_1 = 0$ on a neighborhood of each point of infinity contained in K'. It is easy to see that the function

$$u(p) = g_1(p) + \omega\int_{K'-K_o} N(z, p)\Delta g_1(z)\, dv(z)$$

(dv: the volume element with respect to the local coordinate) for $p \in R'$ is harmonic on R', $u^{K'} = u$ and $u = 0$ on $\partial K_o$, so that $u \equiv 0$. Hence $g_1(p) = -\omega\int_{K'-K_o} N(z, p)\Delta g_1(z)\, dv(z)$. This equation holds also for $p \in \Delta_{\mathcal{N}}$, in particular, for $p = q_1$ and $q_2$. Since $g_1(q_i) = g(q_i)$ $(i = 1, 2)$, $g(q_1) \neq g(q_2)$ implies that there exists $z \in K' - K_o$ such that $N(z, q_1) \neq N(z, q_2)$.

In the above argument, we have also seen that the Kuramochi compactification is metrizable. Furthermore, we can show that it is resolutive with respect to the Dirichlet problem. (See [1] and [7].) These properties of the Kuramochi compactification make it easier to construct a potential theory including the boundary (cf. [1], [6] etc.).

## § 5. Relations with the Martin boundary

There are many similarities between the Kuramochi boundary and the Martin boundary. For example, both are metrizable and resolutive. It is known that if R is a parabolic Riemann surface or a finitely connected plane domain, then its Kuramochi boundary coincides with the Martin boundary. In fact, in the latter case, there exists a one-to-one correspondence among Carathéodory's prime ends, Kuramochi boundary points and Martin boundary points.

On the other hand, Kuramochi [5] gave examples of Riemann surfaces for which there is no homeomorphism between the Kuramochi compactification and the Martin compactification. Although a kernel (usually denoted by K) is considered also for the Martin boundary, it fails to have properties of Green function when its pole is on the boundary, so that it is difficult to construct a potential theory on the Martin compactification. In this sense, we may say that the notion of Kuramochi boundary is essentially different from that of Martin boundary.

### *References*

[1] C. Constantinescu and A. Cornea, Ideale Ränder Riemannscher Flächen, Berlin-Göttingen-Heidelberg, 1963.

[2] J. L. Kelley, General topology, New York, 1955.

[3] Z. Kuramochi, Mass distributions on the ideal boundaries of abstract Riemann surfaces, II, Osaka Math. J., 8 (1956), 145-186.

[4] Z. Kuramochi, Potentials on Riemann surfaces, J. Fac. Sci. Hokkaido Univ. Ser. I, 16 (1962), 5-79.

[5] Z. Kuramochi, Relations among topologies on Riemann surfaces, I-IV, Proc. Japan Acad., 38 (1962), 310-315 and 457-472.

[6] Z. Kuramochi, On boundaries of Riemann surfaces (Japanese), Sûgaku, 16 (1964), 80-94.

[7] F-Y. Maeda, Notes on Green lines and Kuramochi boundary, J. Sci. Hiroshima Univ. Ser. A-I Math., 28 (1964), 59-66.

[8] M. Ohtsuka, An elementary introduction of Kuramochi boundary, Ibid., 271-299.

Department of Mathematics,
Faculty of Science,
Hiroshima University

# II. ON FULL-SUPERHARMONIC FUNCTIONS

Fumi-Yuki MAEDA

## Introduction

As a base space, we consider a non-compact space $\Omega$ of type $\mathcal{E}$ in the sense of Brelot-Choquet [3]. If its dimension $\tau$ is equal to two, then we may understand that $\Omega$ represents an open Riemann surface. Full-superharmonic functions on $\Omega$ are, roughly speaking, functions which are superharmonic on the ideal boundary as well as inside $\Omega$.

The aim of this monograph is to obtain an integral representation of full-superharmonic functions analogous to the Riesz representation of superharmonic functions. As is to be seen, parts of the measures for this representation are forced to be distributed on the ideal boundary and it becomes necessary to realize the ideal boundary in a suitable way. We shall show, by using Choquet's theorem, that the Kuramochi boundary serves our purpose.

On a Riemann surface, such an integral representation has been established in [6] and [5], but our construction of the theory is different from theirs.

## § 1. BLD-functions and Dirichlet principle

To define the concept of full-superharmonic functions, we need the Dirichlet principle. We shall state it in terms of BLD-functions (or, Dirichlet functions), a theory of which is found in [1] or in [5]. Given an open set $\omega$ in $\Omega$, the linear space of all BLD-functions on $\omega$ will be denoted by $D(\omega)$. We assume that $f \in D(\omega)$ always takes the values of normal extension at points at infinity in $\omega$ (cf. [1], n° 24). For $f_1, f_2 \in D(\omega)$, let $\langle f_1, f_2\rangle_\omega$ be the mutual Dirichlet integral over $\omega$ -{points at infinity}. We define $\|f\|_\omega = \langle f, f\rangle_\omega^{1/2}$. Let $C_D^\infty(\omega)$ be the subspace of $D(\omega)$ consisting of functions f on $\omega$ with the following properties: f is infinitely differentiable on $\omega$-{points at infinity} and the support of f is compact in $\omega$. Let $D_0(\omega)$ be the set of all $f \in D(\omega)$ for which there exists a sequence $\{f_n\}$ of functions in $C_D^\infty(\omega)$ such that $\|f_n - f\|_\omega \to 0$ and $f_n \to f$ q.p. on $\Omega$ as $n \to \infty$. Here "q.p." means "except on a polar set". The linear space $D(\omega)$ is the direct sum of two subspaces $HD(\omega)$ and $D_0(\omega)$, where $HD(\omega)$ is the space of harmonic functions h on $\omega$ such that $\|h\|_\omega < \infty$. $HD(\omega)$ and $D_0(\omega)$ are orthogonal to each other with respect to $< \, , \, >_\omega$. For $f \in D(\omega)$, the expression $f = h + f_0$ with $h \in HD(\omega)$ and $f_0 \in D_0(\omega)$ is called the Royden decomposition of f on $\omega$. For a non-polar compact set K in $\Omega$, let $D^K = \{f \in D(\Omega);\ f = 0 \text{ q.p. on } K\}$.

The relative boundary $\partial K$ of a subset K with non-empty interior will be called *piecewise smooth* if it is a regular surface (or curve) in the sense of [3]. In this case, we generally denote by dS the surface (or line) element of $\partial K$ and by $\frac{\partial}{\partial \nu}$ the outer normal derivative on $\partial K$.

*Theorem* 1. (Dirichlet Principle) *Given* $f \in D(\Omega)$ *and a non-polar compact set* K, *there exists a unique* $f^K \in D(\Omega)$ *such that*

(i) $f^K$ *is harmonic on* $\Omega - K$ *and* $f = f^K$ *on* K;

(ii) $\|f^K\|_\Omega = \min\{\|g\|_\Omega;\ g - f \in D^K\}$.

*Furthermore,* $f^K$ *has the following properties:*

(a) *If* $f_1 = f_2$ *q.p. on* K, *then* $f_1^K = f_2^K$ *q.p. on* $\Omega$.

(b) $f = f^K$ *on* $\Omega - K$ *if and only if* $<f, g>_\Omega = 0$ *for all* $g \in D^K$.

(c). *If* K' *is another compact set such that* $K' \supseteq K$, *then* $f^K = (f^{K'})^K = (f^K)^{K'}$.

(d) $f \to f^K$ *is a linear mapping of* $D(\Omega)$ *into itself.*

(e) *If* $f \equiv c$ (constant), *then* $f^K \equiv c$.

(f) *If* $f \geqq 0$, *then* $f^K \geqq 0$.

(g) *If* $\omega$ *is a component of* $\Omega$ - K, *then* $f^K = f^{\partial\omega}$ *on* $\omega$.

(h) *If* $f \geqq 0$ *on* $\partial K$, *then* $f^K \geqq H_f^{\Omega-K}$, *where* $H_f^{\Omega-K}$ *is the Dirichlet solution on* $\Omega$ - K *with boundary conditions* f *on* $\partial K$ *and* 0 *on the ideal boundary.*

(k) *If* K' *is a compact set in* $\Omega$ *containing* K *in its interior and if* $\partial K'$ *is piecewise smooth, then* $\int_{\partial K'} \frac{\partial f^K}{\partial \nu} dS = 0.$

This theorem can be proved by the same method as Satz 15.1 of [5].

If f is a BLD-function on a neighborhood of K, then there exists $f_1 \in D(\Omega)$ such that $f = f_1$ on a neighborhood of K. We define $f^K = f_1^K$. By (a) of the above theorem, $f^K$ is well-defined.

*Corollary. Let* $\omega$ *be a domain in* $\Omega$ *with non-polar compact relative boundary* $\partial\omega$. *If* $b \in \partial\omega$ *is a regular point of* $\omega$ *and if* $f \in D(\Omega)$ *is bounded on* $\partial\omega$ *and continuous at* b, *then* $\lim_{p\to b, p\in\omega} f^{\partial\omega}(p) = f(b)$.

*Proof:* Let $\alpha = \sup_{\partial\omega} f$ and $\beta = \inf_{\partial\omega} f$. By (d), (e), (g) and (h) of the theorem, we have

$$\alpha - f^{\partial\omega} \geqq H^{\omega}_{\alpha-f} \qquad \text{and} \qquad f^{\partial\omega} - \beta \geqq H_{f-\beta}.$$

Since b is regular, $H^{\omega}_{\alpha-f}(p) \to \alpha - f(b)$ and $H^{\omega}_{f-\beta}(p) \to f(b) - \beta$ $(p \to b)$. Hence

$$f(b) = \lim_{p\to b} H^{\omega}_{f-\beta}(p) + \beta \leqq \underline{\lim}_{p\to b}\, f^{\partial\omega}(p)$$

$$\leqq \overline{\lim}_{p\to b}\, f^{\partial\omega}(p) \leqq -\lim_{p\to b} H^{\omega}_{\alpha-f}(p) + \alpha = f(b).$$

Thus, we have the corollary.

## § 2. Full-harmonic measures

Let $\omega$ be an open subset of $\Omega$ such that its relative boundary $\partial\omega$ is compact. A harmonic function h on $\omega$ will be called *full-harmonic* there, if, for any non-polar compact set K in $\Omega$ containing $\partial\omega$ in its interior, $h^{\partial K\cap\omega} = h$ on $\omega$ - K. For example, $f^K$ is full-harmonic on $\Omega$ - K.

In the sequel, we fix a subdomain R of $\Omega$ such that it is not relatively compact and its relative boundary $\partial R$ is non-polar and compact. A compact set in $\Omega$ will be called *admissible* if it contains $\partial R$ in its interior.

For an admissible compact set K, let $C^{\infty}(\partial K \cap R)$ be the space of restrictions of functions in $C_D^{\infty}(R)$ onto $\partial K \cap R$. For $\phi \in C^{\infty}(\partial K \cap R)$, choose $f \in C_D^{\infty}(R)$ such that $f|_{\partial K \cap R} = \phi$ and define $\phi^K = f^{\partial K \cap R}$ on R - K. By (a) of Theorem 1, $\phi^K$ is well-defined. For each $p \in R - K$, the mapping $\phi \to \phi^K(p)$ is a non-negative linear functional on $C^{\infty}(\partial K \cap R)$(by (d) and (f) of Theorem 1). By the Stone-Weierstrass theorem, we see that $C^{\infty}(\partial K \cap R)$ is dense in $C(\partial K \cap R)$, the space of all continuous functions on $\partial K \cap R$ with sup-norm. Hence, there exists a Radon measure $\mu_p^K$ on $\partial K \cap R$ such that $\int \phi \, d\mu_p^K = \phi^K(p)$ for all $\phi \in C^{\infty}(\partial K \cap R)$. If we regard $\mu_p^K$ as a measure on R, then $\int f \, d\mu_p^K = f^{\partial K \cap R}(p)$ for $f \in C_D^{\infty}(R)$. By (e) of Theorem 1, we see that the total mass of $\mu_p^K$ is equal to 1. Obviously, a harmonic function h on R is full-harmonic there if and only if $\int h \, d\mu_p^K = h(p)$ for any admissible K and for any $p \in R - K$.

*Lemma* 1. Let K be an admissible compact set and let $\omega$ be any component of R - K. If f is lower semi-continuous on $\partial K \cap R$ and does not assume the value $-\infty$, then

(a) $u(p) = \int f \, d\mu_p^K$ is either $\equiv \infty$ on $\omega$ or full-harmonic on $\omega$;

(b) $\varliminf_{p \to b, p \in \omega} u(p) \geqq f(b)$ for every regular boundary point b of $\omega$;

the equality holds if f is continuous.

*Proof*: 1) The case $f \in C_D^{\infty}(\partial K \cap R)$: Since $u = f^K$ by definition, u is full-harmonic on R and $\lim_{p \to b} u(p) = f(b)$ by the corollary to Theorem 1.

2) The case $f \in C(\partial K \cap R)$: In this case, we can find $f_n \in C_D^{\infty}(\partial K \cap R)$ such that $\{f_n\}$ converges to f uniformly on $\partial K \cap R$. Let $u_n(p) = \int f_n \, d\mu_p^K$

$\equiv f_n^K(p)$. Then $\{u_n\}$ converges to u uniformly on R - K. Hence u is harmonic on $\omega$. Moreover, if K' is any admissible compact set for $\omega$, then

$$(*) \quad u(p) = \lim_{n\to\infty} u_n(p) = \lim_{n\to\infty} \int u_n \, d\mu_p^{K'} = \int u \, d\mu_p^{K'}$$

for $p \in \omega - K'$. Hence u is full-harmonic on $\omega$. As for (b), we have

$$\lim_{p\to b} u(p) = \lim_{n\to\infty} \lim_{p\to b} u_n(p) = \lim_{n\to\infty} f_n(b) = f(b).$$

3) The case f is lower semi-continuous: We can find $f_n \in C(\partial K \cap R)$ such that $f_n \nearrow f$. Hence (*) also holds for $u_n(p) = \int f_n \, d\mu_p^K$ and we have (a). For (b), we have

$$\varliminf_{p\to b} u(p) \geqq \lim_{n\to\infty} \varliminf_{p\to b} u_n(p) = \lim_{n\to\infty} f_n(b) = f(b).$$

§ 3. Full-superharmonic functions

A superharmonic function s on R is called *full-superharmonic* there if, for any admissible compact set K, $s(p) \geqq \int s \, d\mu_p^K$ for all $p \in R - K$.

Obviously, a function h is full-harmonic if and only if h and -h are both full-superharmonic. The following properties are immediate consequences of the definition:

1) If $s_1$ and $s_2$ are full-superharmonic and $\alpha_1, \alpha_2 > 0$, then $\alpha_1 s_1 + \alpha_2 s_2$ and $\min(s_1, s_2)$ are full-superharmonic.

2) If $s_n$ $(n = 1, 2, \ldots)$ are full-superharmonic (resp. full-harmonic) and $s_n \nearrow s$, $s \not\equiv \infty$, then s is full-superharmonic (resp. full-harmonic).

*Lemma* 2. (Minimum Principle) If s is full-superharmonic on R, bounded below near $\partial R$ and if $\varliminf_{p\to b\in\partial R} s(p) \geqq 0$ q.p. on $\partial R$, then $s \geqq 0$ on R.

*Proof*: Let K be an admissible compact set and let $\alpha = \inf_{p \in \partial K \cap R} s(p)$. From our assumption, it follows that $s \geqq \min(\alpha, 0)$ on $K \cap R$. On the other hand, $s(p) \geqq \int s \, d\mu_p^K \geqq \alpha$ for all $p \in R - K$. Hence $s \geqq \min(\alpha, 0)$ on R. If $\alpha < 0$, then s assumes its minimum on $\partial K \cap R$, so that $s \equiv \text{const.} = \alpha$, which is impossible. Hence $\alpha \geqq 0$, i.e., $s \geqq 0$.

*Lemma* 3. Let s be a superharmonic function on R. Suppose $\{\Omega_n\}$ is an exhaustion of $\Omega$ such that each $\bar{\Omega}_n$ is admissible. If, for each n, $s(p) \geqq \int s \, d\mu_p^{\bar{\Omega}_n}$ for all $p \in R - \bar{\Omega}_n$, then s is full-superharmonic on R.

*Proof*: Let K be any admissible compact set and let f be any continuous function on $\partial K \cap R$ such that $f \leqq s$ there. Let

$$u_f(p) = s(p) - \int f \, d\mu_p^K \qquad \text{for } p \in R - K.$$

We choose an n such that $\Omega_n \supsetneq K$ and fix it. Let $\alpha = \inf_{p \in \partial\Omega_n \cap R} u_f(p)$. By Lemma 1, (a), $v_f(p) = \int f \, d\mu_p^K$ is full-harmonic on R - K, i.e., $\int v_f \, d\mu_p^{\bar{\Omega}_n} = v_f(p)$ for $p \in R - \bar{\Omega}_n$. Hence, using our assumption, we have

$$\alpha \leqq \int u_f \, d\mu_p^{\bar{\Omega}_n} = \int s \, d\mu_p^{\bar{\Omega}_n} - \int v_f \, d\mu_p^{\bar{\Omega}_n} \leqq s(p) - v_f(p) = u_f(p)$$

for any $p \in R - \bar{\Omega}_n$. For any component $\omega$ of R - K and for any regular boundary point b of $\omega$, Lemma 1, (b) implies that $\lim_{p \to b} \int f \, d\mu_p^K = f(b)$. Hence $\underline{\lim}_{p \to b} u_f(p) \geqq s(b) - f(b) \geqq 0$. If $b \in \partial\Omega_n \cap \omega$, then $\underline{\lim}_{p \to b} u_f(p) \geqq u_f(b) \geqq \alpha$. Since $u_f$ is superharmonic on $\Omega_n \cap \omega$, it follows that $u_f \geqq \min(\alpha, 0)$ on $\bar{\Omega}_n \cap \omega$. Hence $u_f \geqq \min(\alpha, 0)$ on $(R - K) \cap \Omega_n$. Thus we have proved that $u_f \geqq \min(\alpha, 0)$ on R - K. By minimum principle for superharmonic functions, $\alpha$ can not be negative. Hence $u_f \geqq 0$ on $R - K$, i.e., $s(p) \geqq \int f \, d\mu_p^K$ for $p \in R - K$. Since $\sup_{f \leqq s} f = s$, $s(p) \geqq \int s \, d\mu_p^K$. Therefore, s is full-superharmonic.

*Corollary.* If s is superharmonic on R, full-superharmonic on R - K for some admissible compact set K, then s is full-superharmonic on R.

*Proof*: If $\{\Omega_n\}$ is an exhaustion of $\Omega$ such that each $\overline{\Omega}_n$ is admissible and $\Omega_1 \supset K$, then $\int s \, d\mu_p^{\overline{\Omega}_n} \leqq s(p)$ for all $p \in R - \overline{\Omega}_n$. Hence s is full-superharmonic by the above lemma.

*Lemma 4.* Given a full-superharmonic function s on R and an admissible compact set K such that every point on $\partial K$ is regular for R - K, then

$$s_K(p) = \begin{cases} s(p) & \text{if } p \in R \cap K \\ \int s \, d\mu_p^K & \text{if } p \in R - K \end{cases}$$

is full-superharmonic on R.

*Proof*: By Lemma 1, (b), we see that $s_K$ is lower semicontinuous on R. Then the definition of full-superharmonic functions implies that $s_K$ is superharmonic on R. Hence, $s_K$ is full-superharmonic by the above corollary and Lemma 1, (a).

## § 4. Full-superharmonic functions of potential type

*Theorem 2. Let* s *be a non-negative full-superharmonic function on* R. *Then there exists the greatest full-harmonic minorant of* s.

*Proof*: We may apply Perron's method with a help of Lemma 4; or we can see by Lemma 1, (a) that $h(p) = \lim_{n \to \infty} \int s \, d\mu_p^{K_n}$ defines the greatest full-harmonic minorant of s, where $\{K_n\}$ is a decreasing sequence of admissible compact sets such that $\bigcap_{n=1}^{\infty} K_n \cap R = \emptyset$.

A non-negative full-superharmonic function on R is said to be *of potential type* if its greatest full-harmonic minorant is zero.

If s is any non-negative full-superharmonic function on R and if h is its greatest full-harmonic minorant, then s - h is a full-superharmonic function of potential type. Conversely, if s = h + w with h full-harmonic and w full-superharmonic of potential type, then h is the greatest full-harmonic minorant of s.

We shall denote by $\mathscr{P} \equiv \mathscr{P}(R)$ the family of all full-superharmonic functions on R of potential type and by $\mathscr{P}_b \equiv \mathscr{P}_b(R)$ the subfamily of $\mathscr{P}$ consisting of functions in $\mathscr{P}$ which are harmonic on R. If s is non-negative full-superharmonic on R and if $s \leq w$ for some $w \in \mathscr{P}$, then $s \in \mathscr{P}$. Thus, $w_1, w_2 \in \mathscr{P}$ implies $\min(w_1, w_2) \in \mathscr{P}$.

*Lemma* 5. Let w be non-negative full-superharmonic on R and harmonic near $\partial R$, i.e., there exists an open set $\omega$ containing $\partial R$ such that w is harmonic on $\omega \cap R$. Then $w \in \mathscr{P}$ if and only if w is bounded near $\partial R$ and $\lim_{p \to b} w(p) = 0$ for every regular boundary point $b \in \partial R$.

*Proof*: Suppose $w \in \mathscr{P}$. We can choose $\omega$ in such a way that $\partial\omega \cap R$ is regular and w is bounded on $\partial\omega \cap R$. Let h be the Dirichlet solution on $\omega \cap R$ with boundary conditions w on $\partial\omega \cap R$ and 0 on $\partial R$. Since $w \geqq 0$ and w is harmonic on $\omega \cap R$, $w \geqq h$ on $\omega \cap R$. Let $h_1 = w - h$. We can see that

$$s = \begin{cases} -h_1 & \text{on } \omega \cap R \\ 0 & \text{on } R - \omega \end{cases}$$

is full-superharmonic (cf. Lemma 4) and $-s \leqq w$ on R. Since $w \in \mathscr{P}$, it follows that $-s \leqq 0$ or $h_1 \leqq 0$ on R. Hence $h_1 = 0$, so that h = w.

Then w is bounded on $\omega \cap R$ and $\lim_{p \to b} w(p) = 0$ for any regular point $b \in \partial R$.

Conversely, suppose w is bounded near $\partial R$ and $\lim_{p \to b} w(p) = 0$ for any regular point $b \in \partial R$. If $h_o$ is a full-harmonic minorant of w, then $\overline{\lim}_{p \to b} h_o(p) \leqq 0$. Hence Lemma 2 implies $h_o \leqq 0$, so that $w \in \mathcal{P}$.

*Corollary.* $\mathcal{P}$ and $\mathcal{P}_b$ are cones.

## § 5. Kuramochi kernel

In order to obtain an integral representation of full-superharmonic functions, we need to choose a suitable kernel. By analogy with Riesz representation for superharmonic functions, we require that our kernel N satisfies the following conditions:

(i) $N(z, p)$ is defined on $R \times R$ taking values in $(0, \infty]$;

(ii) $N(z, p)$ has the same singularity at $z = p$ with the Green function $G(z, p)$ of R, i.e., $N(z, p) - G(z, p)$ is harmonic on R for each $p \in R$;

(iii) For each $p \in R$, $N_p \in \mathcal{P}$ $(N_p(z) = N(z, p))$ and $N_p$ is full-harmonic on $R - \{p\}$.

It is easy to see that if such a kernel exists then it is unique. We shall show the existence by a method of [5].

Let $HD_o = \{u \in HD(R);$ there is $f \in D^{\partial R}$ such that $f|_R = u\}$. If $u \in HD_o$ and if $\omega$ is an open set containing $\partial R$, then it is known that $u = H_v^{\omega \cap R}$ on $\omega \cap R$, where $v = u$ on $\partial\omega \cap \partial R$ and $v = 0$ on $\partial R$ (see [1], Th. 8 and its extension). Hence, in particular, any $u \in HD_o$ is bounded near $\partial R$. It is easy to see that $HD_o$ is a Hilbert space with respect to the inner product $< , >_R$. For each $p \in R$, the mapping $u \to q_\tau u(p)$ ($q_\tau$ is a constant depending only on $\tau$. It is denoted by $\phi_\tau$ in [3].) is a bounded linear form on $HD_o$. (Cf. [5], p. 160.) Hence there exists $U_p \in HD_o$ such that $<U_p, u>_R = q_\tau u(p)$ for any $u \in HD_o$. Since

$U_p(z) = (1/q_\tau)\langle U_z, U_p\rangle_R$, we have $U_p(z) = U_z(p)$ and $U_p(p) \geqq 0$. Now, we define $N_p = G_p + U_p$, where $G_p(z) = G(z, p)$. It is obvious that $N(z, p) = N_p(z)$ satisfies condition (ii). Furthermore, $N(z, p) = N(p, z)$.

*Lemma* 6. If K is an admissible compact set containing p in its interior, then

$$N_p(z) = \int N_p \, d\mu_z^K \qquad \text{for all } z \in R - K.$$

*Proof*: We fix p and K. $G_p$ is bounded on R - K. Thus let $\alpha_o = \sup_{z\in R-K} G_p(z)$. For any $\alpha \geqq \alpha_o$, let $f_\alpha = \min(G_p, \alpha) + U_p$ on R. Since $f_\alpha = N_p$ on $\partial K \cap R$, $\int f_\alpha \, d\mu_z^K = \int N_p \, d\mu_z^K$ $(z \in R - K)$. Let $f = f_{\alpha_o}$ on R, = 0 on $\Omega - R$. Then $f \in D(\Omega)$. It is enough to show that $f^{\partial K\cap R} = f$ on R - K, or, by Theorem 1, (b), that $\langle f, g\rangle_\Omega = 0$ or $\langle f_{\alpha_o}, g\rangle_R = 0$ for all $g \in D^K$. We may assume that g = 0 in a neighborhood of p for each $g \in D^K$. Let $g = u + g_o$ be the Royden decomposition of g on R. Then $u \in HD_o$. Hence

$$\langle f_{\alpha_o}, g\rangle_R = \langle f_\alpha, g\rangle_R = \langle \min(G_p, \alpha), g_o\rangle_R + \langle U_p, u\rangle_R.$$

Now, $\langle \min(G_p, \alpha), g_o\rangle_R \to q_\tau g_o(p)$ as $\alpha \to \infty$, since $g_o$ is continuous in a neighborhood of p. Hence

$$\langle f_{\alpha_o}, g\rangle_R = q_\tau g_o(p) + q_\tau u(p) = q_\tau g(p) = 0.$$

By this lemma, we see that $N_p$ is full-harmonic on $R - \{p\}$. Hence, by the corollary to Lemma 3, $N_p$ is full-superharmonic on R.

*Lemma* 7. $N_p \in \mathcal{P}$ and $U_p \in \mathcal{P}_b$.

*Proof*: Let K be an admissible compact set containing p in its interior. By (h) of Theorem 1, $\int G_p \, d\mu_z^K \geqq G_p(z)$ for $z \in R - K$. Hence it follows from the above lemma that

$$\int U_p \, d\mu_z^K = \int N_p \, d\mu_z^K - \int G_p \, d\mu_z^K \leqq N_p(z) - G_p(z) = U_p(z).$$

Therefore, $U_p$ is full-superharmonic. Since $U_p \in HD_o$, Lemma 2 implies $U_p \geqq 0$. Hence $N_p \geqq G_p > 0$. Since $U_p \in HD_o$ and $\lim_{z \to b} G_p(z) = 0$ for regular point $b \in \partial R$, Lemma 5 implies $N_p \in \mathcal{P}$ and $U_p \in \mathcal{P}_b$.

Thus we have seen that $N(z, p) = N_p(z)$ satisfies the required conditions (i), (ii) and (iii).

*Lemma* 8. Let K be an admissible compact set in $\Omega$ such that $\partial K \cap R$ is piecewise smooth. If $p \notin K$, then $\int_{\partial K \cap R} \frac{\partial N_p}{\partial \nu} dS = q_\tau$; if p is in the interior of K, then the integral = 0.

*Proof*: We choose a spherical neighborhood of p such that $\bar{\omega}_p \cap (\partial K \cup \partial R) = \emptyset$. Let $\omega'$ be another neighborhood of p such that $\bar{\omega}' \subset \omega_p$ and let $\alpha = \sup_{z \in \partial \omega'} N_p(z)$. Let $f = \min(N_p, \alpha)$ on R, = 0 on $\Omega - R$. Then $f \in D(\Omega)$ and $f = N_p$ on $R - \omega'$. Let K' be another admissible compact set such that it is contained in the interior of K and $K' \cap \bar{\omega}_p = \emptyset$. By Lemma 6, we have, for $z \in R - (K' \cup \bar{\omega}')$,

$$f(z) = N_p(z) = \int N_p \, d\mu_z^{K' \cup \bar{\omega}'} = \int f \, d\mu_z^{K' \cup \bar{\omega}'} = f^{K' \cup \bar{\omega}'}(z).$$

If $p \notin K$, then it follows from Theorem 1, (k) that $\int_{\partial \omega_p \cup (\partial K \cap R)} \frac{\partial f}{\partial \nu} dS = 0$. Since $\int_{\partial \omega_p} \frac{\partial N_p}{\partial \nu} dS = -q_\tau$, we have $\int_{\partial K \cap R} \frac{\partial N_p}{\partial \nu} dS = q_\tau$. If p is in the interior of K, then $\int_{\partial K \cap R} \frac{\partial f}{\partial \nu} dS = 0$, i.e., $\int_{\partial K \cap R} \frac{\partial N_p}{\partial \nu} dS = 0$.

*Theorem* 3. *Let* $s \in \mathcal{P}$ *and let* $\nu$ *be the measure associated with* s, i.e., *let* $s = \int G_p \, d\nu(p) + h$ *be the Riesz decomposition of* s. *Then* $\int N_p \, d\nu(p) \in \mathcal{P}$ *and there exists* $w \in \mathcal{P}_b$ *such that* $s = \int N_p \, d\nu(p) + w$.

*Furthermore, if* $s = \int N_p \, d\nu_1(p) + w_1$ *with a harmonic function* $w_1$ *on* R, *then* $\nu_1 = \nu$.

*Proof*: 1) Let $\delta$ be a compact set in R. We can easily see that $v_\delta = \int_\delta N_p \, d\nu(p)$ is non-negative full-superharmonic. Let $w_\delta = -\int_\delta U_p \, d\nu(p) + h + \int_{\Omega_0 - \delta} G_p \, d\nu(p)$. Then $w_\delta$ is superharmonic on R and $s = v_\delta + w_\delta$. Since $v_\delta$ is full-harmonic on R - $\delta$, we see from Lemma 3 that $w_\delta$ is full-superharmonic on R. Since $U_p \in HD_0$, it follows from Harnack's principle that $\lim_{z\to b} U_p(z) = 0$ uniformly for $p \in \delta$, for each regular boundary point $b \in \partial R$. Hence $\lim_{z\to b} \int_\delta U_p(z) d\nu(p) = 0$, so that

$$\underline{\lim}_{z\to b}\, w_\delta(z) \geqq \underline{\lim}_{z\to b}\, h(z) \geqq 0.$$

Since $w_\delta$ is bounded below near $\partial R$, we have $w_\delta \geqq 0$ by Lemma 2, i.e., $v_\delta \leqq s$. Since $v_\delta \nearrow \int N_p \, d\nu(p)$ as $\delta \nearrow R$, it follows that $v(z) = \int N_p(z) d\nu(p)$ is non-negative full-superharmonic on R and $v \leqq s$. On the other hand, $w_\delta$ decreases to a harmonic function w on R. It is easy to see that w is non-negative full-superharmonic on R. Obviously, s = v + w. Since $0 \leqq v \leqq s$, $0 \leqq w \leqq s$ and $s \in \mathcal{P}$, $v \in \mathcal{P}$ and $w \in \mathcal{P}_b$.

2) If $s = \int N_p \, d\nu_1(p) + w_1$ with $w_1$ harmonic, then $\int N_p \, d\nu(p) - \int N_p \, d\nu_1(p)$ is harmonic on R, so that $\int G_p \, d\nu(p) - \int G_p \, d\nu_1(p)$ is harmonic on R. Hence $\nu_1 = \nu$.

## § 6. Representation of $w \in \mathcal{P}_b$

We have seen that any $s \in \mathcal{P}$ can be uniquely decomposed into $s = \int N_p \, d\nu(p) + w$ with a measure $\nu$ on R and $w \in \mathcal{P}_b$. Hence our next task is to obtain an integral representation for $w \in \mathcal{P}_b$.

The measure for such a representation can no longer be distributed inside R. This fact requires us to consider something more than R. Thus we try to use Choquet's representation theorem (see [4]):

*Choquet's theorem*: Let E be a Hausdorff locally convex space over the real field, let X be a metrizable compact convex subset of E and let e(X) be the set of all extreme points of X. Then, for any $x_o \in X$, there exists a unit measure $\nu$ on X such that $\nu(X - e(X)) = 0$ (In this case, we say that $\nu$ is a measure on e(X).) and $x_o = \int x \, d\nu(x)$. If, in addition, X is a base of a cone which is a lattice with respect to the order induced by itself, then $\nu$ is uniquely determined.

To apply this theorem to our case, we take the space H of harmonic functions on R as E. $\mathcal{P}_b$ is a cone in H. The topology of compact convergence is introduced in H. With this topology, H is a metrizable locally convex space. Now, let $K_o$ be an admissible compact set such that $\partial K_o \cap R$ is piecewise smooth and let

$$\mathcal{P}_{b,o} = \left\{w \in \mathcal{P}_b;\ \int_{\partial K_o \cap R} \frac{\partial w}{\partial \nu}\, dS = q_\tau\right\}.$$

The set $\mathcal{P}_{b,o}$ is independent of the choice of $K_o$. $\mathcal{P}_{b,o}$ is a convex set as a base of the cone $\mathcal{P}_b$. We shall show that $\mathcal{P}_{b,o}$ is compact and $\mathcal{P}_b$ is a lattice with respect to the order induced by itself.

*Lemma* 9. $\mathcal{P}_{b,o}$ is compact.

*Proof*: Let $p \in R$. We shall first show that $\{w(p);\ w \in \mathcal{P}_{b,o}\}$ is bounded. Suppose it is not. Then there exist $w_n \in \mathcal{P}_{b,o}$ such that $w_n(p) \nearrow \infty$ $(n \to \infty)$. By Harnack's principle, we have $w_n \to \infty$ uniformly on $\partial K_o \cap R$. On the other hand, if we consider the harmonic measure h on $(K_o - \partial K_o) \cap R$ with boundary values 1 on $\partial K_o \cap R$ and 0 on $\partial R$, then Green's formula implies that

$$\int_{\partial K_o \cap R} \frac{\partial w}{\partial \nu}\, dS = \int_{\partial K_o \cap R} w \frac{\partial h}{\partial \nu}\, dS \qquad \text{for any } w \in \mathcal{P}_b.$$

Since $\frac{\partial h}{\partial \nu} \geqq 0$ on $\partial K_o \cap R$ and $\int_{\partial K_o \cap R} \frac{\partial h}{\partial \nu}\, dS > 0$, we have

$$q_\tau = \int_{\partial K_o \cap R} \frac{\partial w_n}{\partial \nu}\, dS = \int_{\partial K_o \cap R} w_n \frac{\partial h}{\partial \nu}\, dS$$

$$\geqq (\inf_{\partial K_o \cap R} w_n)\left(\int_{\partial K_o \cap R} \frac{\partial h}{\partial \nu}\, dS\right) \to \infty,$$

which is impossible. Hence $\{w(p);\ w \in \mathcal{P}_{b,o}\}$ is bounded. Then it follows that $\mathcal{P}_{b,o}$ is a normal family of harmonic functions on R, i.e., it is relatively compact. If $w_n \to w_o$, $w_n \in \mathcal{P}_{b,o}$, then we easily see that $w_o$ is non-negative harmonic, full-superharmonic and $\int_{\partial K_o \cap R} \frac{\partial w_o}{\partial \nu}\, dS = q_\tau$. Also Lemma 5 implies that $w_o \in \mathcal{P}_b$. Hence $w_o \in \mathcal{P}_{b,o}$ and this completes the proof.

*Lemma* 10. $\mathcal{P}_b$ is a lattice with respect to the order induced by itself.

*Proof*: What we have to show is the following: If $w_1, w_2 \in \mathcal{P}_b$, then there exists $w^* \in \mathcal{P}_b$ having the following two properties: (i) $w^* - w_i \in \mathcal{P}_b (i = 1, 2)$; (ii) if $u \in \mathcal{P}_b$ satisfies $u - w_i \in \mathcal{P}_b$ $(i = 1, 2)$, then $u - w^* \in \mathcal{P}_b$.

Now, let $\mathcal{U} = \{s \in \mathcal{P}\ ;\ s - w_i \in \mathcal{P}\ (i = 1, 2)\}$. Since $w_1 + w_2 \in \mathcal{U}$, $\mathcal{U}$ is non-empty. Let $w^* = \inf \mathcal{U}$. Obviously, $w^* \geqq w_i$ $(i = 1, 2)$. We set $w_o \equiv 0$ and for each $s \in \mathcal{U}$ we consider $s_i = s - w_i$ $(i = 0, 1, 2)$. Then $s_i \in \mathcal{P}$ and $w^* - w_i = \inf_{s \in \mathcal{U}} s_i$ $(i = 0, 1, 2)$. Let $\mathcal{U}_i = \{s_i;\ s \in \mathcal{U}\}$ $(i = 0, 1, 2;\ \mathcal{U}_o \equiv \mathcal{U})$.

Since $\mathcal{U}$ is closed under min. operation, so are $\mathcal{U}_i$. It is easy to see that each $\mathcal{U}_i$ is a Perron's family on R in the sense of [5], so that $w^* - w_i$ is harmonic on R for each i = 0, 1, 2. Let K be any admissible compact set. Given $\varepsilon > 0$ and $p \in R - K$, there exists $s \in \mathcal{U}$ such that $w^*(p) > s(p) - \varepsilon$. Then

$$w^*(p) - w_i(p) > s_i(p) - \varepsilon$$

$$\geqq \int s_i \, d\mu_p^K - \varepsilon \geqq \int (w^* - w_i) \, d\mu_p^K - \varepsilon$$

(i = 0, 1, 2). Since $\varepsilon$ is arbitrary, $w^*(p) - w_i(p) \geqq \int (w^* - w_i) d\mu_p^K$ (i = 0, 1, 2). Therefore, $w^* - w_i$ are full-superharmonic. Since $0 \leqq w^* - w_i \leqq s_i \in \mathcal{P}$, $w^* - w_i \in \mathcal{P}_b$ (i = 0, 1, 2). Thus we have seen that $w^* \in \mathcal{P}_b$ and $w^*$ satisfies the property (i). To show the property (ii), let $u \in \mathcal{P}_b$ satisfy $u - w_i \in \mathcal{P}_b$ (i = 1, 2). Obviously, $u \geqq w^*$, so that $u - w^*$ is non-negative harmonic on R. Let K be an admissible compact set such that $\partial K \cap R$ is regular (with respect to the Dirichlet problem on R - K). Let $f(p) = \int (u - w^*) d\mu_p^K$ for $p \in R - K$ and let

$$g = \begin{cases} \inf (u - f, w^*) & \text{on } R - K \\ w^* & \text{on } K \cap R. \end{cases}$$

Then g is continuous on R. Since f is full-harmonic on R - K and $w^*$ is full-superharmonic on R, it follows that g is full-superharmonic and non-negative on R (Corollary to Lemma 3). By replacing u by $u - w_1$ and $w^*$ by $w^* - w_1$, we similarly see that $g - w_1$ and $g - w_2$ are non-negative full-superharmonic on R. Since $0 \leqq g - w_i \leqq w^* - w_i \in \mathcal{P}_b$, we have $g - w_i \in \mathcal{P}$ (i = 0, 1, 2), so that $g \in \mathcal{U}$. Therefore $g = w^*$. It follows then that $u - f \geqq w^*$ on R - K. Thus we conclude that $u - w^*$ is full-superharmonic on R. Since $0 \leqq u - w^* \leqq u \in \mathcal{P}_b$, $u - w^* \in \mathcal{P}_b$.

Now we can apply Choquet's theorem with $E = H$ and $X = \mathcal{P}_{b,o}$ and we obtain

*Theorem 4. For each* $w \in \mathcal{P}_{b,o}$, *there exists a uniquely determined unit measure* $\nu$ *on* $e(\mathcal{P}_{b,o})$ *such that*

$$(**) \qquad w = \int_{\mathcal{P}_{b,o}} v \, d\nu(v).$$

*Corollary. For each* $w \in \mathcal{P}_b$, *there exists a uniquely determined measure* $\nu$ *on* $e(\mathcal{P}_{b,o})$ *such that* (**) *holds.*

§ 7. Realization of $e(\mathcal{P}_{b,o})$ as a part of the Kuramochi boundary

The set $e(\mathcal{P}_{b,o})$ is a family of functions and has no apparent connection with the base space $\Omega$ or R. Our next question is to investigate if $e(\mathcal{P}_{b,o})$ can be identified with any kind of ideal boundary of $\Omega$ or of R. The kernel N for our integral representation suggests that this ideal boundary would be the Kuramochi boundary. We shall see that this is actually the case.

Let $K_1$ be an admissible compact set containing $K_o$ in its interior and let $\mathcal{N}_1 = \{N_p;\ p \in R - (K_1 - \partial K_1)\}$. We consider a metric d on $\mathcal{N}_1$ defined by

$$d(N_{p_1}, N_{p_2}) = \sup_{z \in K_o \cap R} \left| \frac{N_{p_1}(z)}{1+N_{p_1}(z)} - \frac{N_{p_2}(z)}{1+N_{p_2}(z)} \right|$$

The mapping $p \to N_p$ is a homeomorphism of $R - (K_1 - \partial K_1)$ onto $\mathcal{N}_1$ with this metric. Let $\hat{\mathcal{N}}_1$ be the completion of $\mathcal{N}_1$ with respect to d. It is easy to see that $\hat{\mathcal{N}}_1$ is compact (cf. [8]).

*Lemma* 11. $e(\mathcal{P}_{b,o}) \subseteq \hat{\mathcal{N}}_1 - \mathcal{N}_1 \subseteq \mathcal{P}_{b,o}$.

*Proof*: If $w \in \hat{\mathcal{N}}_1 - \mathcal{N}_1$, then there exists a Cauchy sequence $\{N_{p_n}\}$ defining w. Then $\{p_n\}$ has no limit point in R, so that $\{N_{p_n}\}$

converges uniformly on each compact set in R. w is identified with $\lim_{n\to\infty} N_{p_n}$. Then w is harmonic and full-superharmonic on R. Since the convergence is locally uniform, Lemmas 7 and 5 imply that $w \in \mathcal{P}_b$. By Lemma 8, we see that $w \in \mathcal{P}_{b,o}$. Hence $\hat{\mathcal{N}}_1 - \mathcal{N}_1 \subseteq \mathcal{P}_{b,o}$.

Next, let $\{\Omega_n\}$ be an exhaustion of $\Omega$ such that $\Omega_1 \supset K_1$ and each $\partial\Omega_n \cap R$ is piecewise smooth. Given $w \in e(\mathcal{P}_{b,o})$, let $s_n = w_{\bar{\Omega}_n}$ (cf. Lemma 4). Since $0 \leqq s_n < w \in \mathcal{P}_b$, $s_n \in \mathcal{P}$. Let $s_n = \int N_p \, d\nu_n(p) + w_n$ be the decomposition of $s_n$ given in Theorem 3, i.e., $\nu_n$ is the measure associated with $s_n$ and $w_n \in \mathcal{P}_b$. The measure $\nu_n$ is supported by $\partial\Omega_n \cap R$. Since $s_n$ is full-harmonic on $R - \bar{\Omega}_n$, $w_n$ is full-harmonic on R. Hence $0 \leqq w_n \in \mathcal{P}_b$ implies $w_n = 0$. Now, by Lemma 8,

$$\int_{\partial\Omega_m \cap R} \frac{\partial s_n}{\partial\nu} \, dS = 0 \qquad \text{for } m > n.$$

Hence

$$\nu_n(\partial\Omega_n) = \frac{1}{q_\tau}\left(\int_{\partial K_o \cap R} \frac{\partial s_n}{\partial\nu} \, dS - \int_{\partial\Omega_m \cap R} \frac{\partial s_n}{\partial\nu} \, dS\right)$$

$$= \frac{1}{q_\tau} \int_{\partial K_o \cap R} \frac{\partial s_n}{\partial\nu} \, dS = \frac{1}{q_\tau} \int_{\partial K_o \cap R} \frac{\partial w}{\partial\nu} \, dS = 1.$$

If we regard $\nu_n$ as a measure on $\mathcal{N}_1 \subset \hat{\mathcal{N}}_1$ by the homeomorphism $p \to N_p$, then $\{\nu_n\}$ is a sequence of measures on a compact space $\hat{\mathcal{N}}_1$ with bounded total masses. Hence there exists a subsequence $\{\nu_{n_k}\}$ vaguely converging to a measure $\nu_o$ on $\hat{\mathcal{N}}_1$. Then for each $z \in R$,

$$w(z) = \lim_{n\to\infty} s_n(z) = \lim_{k\to\infty} \int N_p(z)\, d\nu_{n_k}(p) = \int_{\hat{\mathcal{N}}_1} v(z)\, d\nu_o(v).$$

Since w is harmonic on R, $\nu_o$ has no mass in $\mathcal{N}_1$. Hence $w = \int_{\hat{\mathcal{N}}_1 - \mathcal{N}_1} v\, d\nu_o(v)$. Since $\hat{\mathcal{N}}_1 - \mathcal{N}_1 \subseteq \mathcal{P}_{b,o}$ and since $w \in e(\mathcal{P}_{b,o})$, it follows that $\nu_o$ is a point mass on $\hat{\mathcal{N}}_1 - \mathcal{N}_1$, so that $w \in \hat{\mathcal{N}}_1 - \mathcal{N}_1$.

Finally, we shall see that $\hat{\mathcal{N}}_1 - \mathcal{N}_1$ can be identified with a part of the Kuramochi boundary of $\Omega$. We follow [5] for a definition of the Kuramochi compactification. (Also see [7] and [8].)

*Lemma* 12. Let $\hat{\Omega}_N$ be the Kuramochi compactification of $\Omega$ and let $\Delta_N = \hat{\Omega}_N - \Omega$ be the Kuramochi boundary of $\Omega$. Let $\hat{R}$ be the closure of R in $\hat{\Omega}_N$. Then the homeomorphism $p \to N_p$ of $R - (K_1 - \partial K_1)$ onto $\mathcal{N}_1$ can be extended to a homeomorphism of $\hat{R}$ onto $\hat{\mathcal{N}}_1$, so that $\hat{R} \cap \Delta_N$ and $\hat{\mathcal{N}}_1 - \mathcal{N}_1$ are topologically equivalent.

*Proof*: Let $\mathcal{N} = \{f;$ continuous on $\Omega$ and $f = f^K$ for some compact set $K\}$. Then $\hat{\Omega}_N$ is the $\mathcal{N}$-compactification of $\Omega$ ([5], [7], [8]). For each $z \in R$, there exists $f_z \in \mathcal{N}$ such that $f_z = N_z$ on $R - K$ and $f_z = 0$ on $\Omega - R - K$ for some (admissible) compact set K. As $p_i \to \xi \in \hat{R} \cap \Delta_N$, $p_i \in R$, $f_z(p_i) \to f_z(\xi)$ for each z. Hence $\{N_{p_i}\}$ becomes a Cauchy sequence in $\mathcal{N}_1$ so that $N_{p_i} \to w_\xi \in \hat{\mathcal{N}}_1 - \mathcal{N}_1$. Since $w_\xi(z) = f_z(\xi)$, the mapping $\xi \to w_\xi$ is well-defined. On the other hand, we can show that $\{f_z;\ z \in R\}$ separates points of $\hat{R} \cap \Delta_N$ (cf. [5], p. 170 or [8]). Then it follows that $N_p \to w \in \hat{\mathcal{N}}_1 - \mathcal{N}_1$ implies $p \to \xi'$ for some $\xi' \in \hat{R} \cap \Delta_N$. Thus we conclude that the mapping $\xi \to w_\xi$ is a continuous extension of the mapping $p \to N_p$ over $\hat{R}$ and it is a homeomorphism of $\hat{R} \cap \Delta_N$ onto $\hat{\mathfrak{N}}_1 - \mathfrak{N}_1$.

Let $\Delta_{1,R}$ be the subset of $\hat{R} \cap \Delta_N$ corresponding to $e(\mathfrak{P}_{b,o})$. ($\xi \in \Delta_{1,R}$ is called a minimal point on $\hat{R} \cap \Delta_N$.) Then Theorem 4 can be restated in the following form with helps of Lemmas 11 and 12:

*Theorem 4'.* *For any* $w \in \mathfrak{P}_b$, *there exists a uniquely determined measure* $\nu$ *on* $\Delta_{1,R} \subseteq \hat{R} \cap \Delta_N$ *such that*

$$w = \int_{\Delta_{1,R}} N_\xi \, d\nu(\xi),$$

*where* $N_\xi(z) = \lim_{p \to \xi} N_p(z)$ *for* $\xi \in \hat{R} \cap \Delta_N$ *and* $z \in R$.

Combining this theorem with Theorems 2 and 3, we have our final theorem:

*Theorem 5.* *For any non-negative full-superharmonic function* s, *there exists a measure* $\nu$ *on* $\hat{R}$ *such that*

$$s = \int_{\hat{R}} N_p \, d\nu(p) + h,$$

*where* h *is full-harmonic;* $\nu$ *is uniquely determined as a measure on* $R \cup \Delta_{1,R}$.

*References*

[1] M. Brelot, Étude et extensions du principe de Dirichlet, Ann. Inst. Fourier, 5 (1955), 371-419.

[2] M. Brelot, Axiomatique des fonctions harmoniques et surharmoniques dans un espace localement compact, Sém. Théorie Pot., 2 (1958), no. 1, 40 pp.

[3] M. Brelot and G. Choquet, Expaces et lignes de Green, Ann. Inst. Fourier, 3 (1952), 199-263.

[4] G. Choquet, Existence et unicité des représentations intégrales au moyen des points extrémaux dans les cones convexes, Sém. Bourbaki, 9 (1956-57), no. 139, 15 pp.

[5] C. Constantinescu and A. Cornea, Ideale Ränder Riemannscher Flächen, Berlin-Göttingen-Heidelberg, 1963.

[6] Z. Kuramochi, Potentials on Riemann surfaces, J. Fac. Sci. Hokkaido Univ. Ser. I, 16 (1962), 5-79.

[7] F-Y. Maeda, Notes on Green lines and Kuramochi boundary of a Green space, J. Sci. Hiroshima Univ. Ser. A-I Math., 28 (1964), 59-66.

[8] F-Y. Maeda, Introduction to the Kuramochi boundary, these proceedings, No. 1.

Department of Mathematics,
Faculty of Science,
Hiroshima University

# III. RIEMANN SURFACES WITH MARTIN AND KURAMOCHI BOUNDARY POINTS

Hiroshi TANAKA

## Introduction

There are two typical theories of ideal boundaries of an open Riemann surface R, Martin's and Kuramochi's. The purpose of this report is to construct a compactification of R with ideal boundary of mixed type and to investigate the correspondence between the new ideal boundary and the Martin boundary or the Kuramochi boundary.

Let $\Delta$ be the ideal boundary of R in the sense of the Kerékjártó-Stoilow compactification of R and let A be a non-empty closed or relatively open subset of $\Delta$ such that $A \neq \Delta$. We construct a kernel function $K^A(P, Q)$ which has the boundary behavior like the Martin kernel near A and like the Kuramochi kernel near $\Delta - A$. We can consider the compactification of R with respect to $K^A(P, Q)$ and denote its ideal boundary by $\Delta_{K^A}$.

Our main results are as follows:

Theorem 3. If A is closed, then the part of $\Delta_{K^A}$ lying on $\Delta - A$ is homeomorphic to the part of the Kuramochi boundary lying on $\Delta - A$.

Theorem 4. If A is relatively open, then the part of $\Delta_{K^A}$ lying on A is homeomorphic to the part of the Martin boundary lying on A.

## § 1. Preliminaries

Let R be an open Riemann surface. We shall call a domain on R an *end* if it is not relatively compact and its relative boundary in R is not empty and consists of a finite number of closed analytic curves. In this report, an *approximation* of R will mean a sequence

$\{\Omega_n\}$ of ends on R such that $\Omega_n \cup \partial\Omega_n \subset \Omega_{n+1}$ (n = 1, 2,...) and $\bigcup_n \Omega_n = R$, where $\partial\Omega_n$ means the relative boundary of $\Omega_n$ in R.

Let $\Delta$ be the ideal boundary of R in the Kerékjártó-Stoilow compactification of R. Let E be any subset of R. We shall denote by B(E) the interesection of $\Delta$ and the closure of E in $R \cup \Delta$. We shall say that a subset of $\Delta$ is *isolated* if it is relatively open and closed in $\Delta$. For any non-empty subset A of $\Delta$ such that $A \neq \Delta$, it is isolated if and only if there exists an end $\Omega$ such that $B(\Omega) = A$. Let A be a non-empty closed subset of $\Delta$ such that $A \neq \Delta$. Then there exists an approximation $\{\Omega_n\}$ of R such that $B(R - \Omega_n) \searrow A$ as $n \to \infty$, i.e. $B(\Omega_n) \nearrow \Delta - A$ as $n \to \infty$. If A is isolated, then we can choose $\{\Omega_n\}$ such that $B(\Omega_n) = \Delta - A$ for every n.

## § 2. Dirichlet Principle

Let G be an open set in R. A continuous function f in G will be called *piecewise smooth* (cf. [3]) if it is continuously differentiable in an open subset G' of G such that G - G' locally consists of a finite number of points and open analytic arcs. If the mixed Dirichlet integral

$$\iint_G \left( \frac{\partial f_1}{\partial x} \frac{\partial f_2}{\partial x} + \frac{\partial f_1}{\partial y} \frac{\partial f_2}{\partial y} \right) dxdy$$

exists for piecewise smooth functions $f_1$ and $f_2$, it will be denoted by $(f_1, f_2)_G$. The notation $\|f\|_G$ will be used for $(f, f)_G^{1/2}$ and $\|f\|_G$ will be called the *Dirichlet norm* of f or the *norm* of f.

We shall call a compact set in R *regular* if its boundary consists of a finite number of analytic arcs. Let K be a regular compact set in R and let $\phi$ be a continuous function on $\partial K$. We shall denote by $\mathcal{D}(\phi, K)$ the class of all piecewise smooth Dirichlet finite (i.e. with finite Dirichlet norm) functions in R - K with boundary values

$\phi$ on $\partial K$.

2.1 Discussion on ends.

Let $\Omega$ be an end on R and let K be a regular compact set in $\Omega$. Given a continuous function $\phi$ on $\partial K$, we shall denote by $\mathcal{D}(\phi, K, \Omega)$ the subclass of $\mathcal{D}(\phi, K)$ consisting of functions which vanish outside $\Omega$.

Then we can prove (Cf. Theorem 1 in [3])

*Lemma* 1. Suppose $\mathcal{D}(\phi, K, \Omega) \neq \emptyset$. Then there exists a uniquely determined function $h \in \mathcal{D}(\phi, K, \Omega)$ such that $(f - h, h)_{\Omega-K} = 0$ for any $f \in \mathcal{D}(\phi, K, \Omega)$. Furthermore, h is harmonic in $\Omega - K$.

We see that h has the smallest Dirichlet norm among the functions in $\mathcal{D}(\phi, K, \Omega)$. We shall denote $h(P)$ by $\phi_K^\Omega(P)$. If $\phi \equiv 1$ on $\partial K$, then $\phi_K^\Omega(P)$ will be denoted by $\omega(P; K, \Omega)$.

The following properties can be proved easily (Cf. [3]): Let $\phi$, $\phi_i$ $(i = 1, 2)$ be given continuous functions on $\partial K$ such that $\mathcal{D}(\phi, K, \Omega) \neq \emptyset$ and $\mathcal{D}(\phi_i, K, \Omega) \neq \emptyset$ $(i = 1, 2)$.

1) $(a_1\phi_1 + a_2\phi_2)_K^\Omega = a_1(\phi_1)_K^\Omega + a_2(\phi_2)_K^\Omega$ for any real numbers $a_1$, $a_2$.

2) $0 < \omega(P; K, \Omega) \leqq 1$ for $P \in \Omega - K$.

3) If $\phi \geqq 0$ on $\partial K$, then $\phi_K^\Omega \geqq 0$ on $\Omega - K$.

4) If K, K' are regular compact sets in $\Omega$ such that $K \subset K'$, then $(\phi_K^\Omega)_{K'}^\Omega(P) = \phi_K^\Omega(P)$ for $P \in \Omega - K'$.

By 1), 2) and 3), we have the maximum principle for $\phi_K^\Omega$:

5) $|\phi_K^\Omega(P)| \leqq \max_{\partial K} |\phi|$.

2.2 A generalization of the Dirichlet principle.

Hereafter, we shall always assume that A *is a non-empty closed or relatively open subset of* $\Delta$ *such that* $A \neq \Delta$. Let K be a regular compact set in R and let $\phi$ be a given continuous function on $\partial K$.

First we suppose that A is closed. We choose an approximation $\{\Omega_n\}$ of R such that $B(R - \Omega_n) \searrow A$ as $n \to \infty$. Since the closure of $\bigcup_{n=1}^{\infty} \mathcal{D}(\phi, K, \Omega_n)$ in $\mathcal{D}(\phi, K)$ with respect to the Dirichlet norm does not depend on the choice of $\{\Omega_n\}$, we denote it by $\mathcal{D}^A(\phi, K)$.

If A is relatively open, then there is a sequence $\{A_n\}$ of non-empty isolated subsets of $\Delta$ such that $A_n \nearrow A$ as $n \to \infty$. Evidently $\mathcal{D}^{A_n}(\phi, K) \supset \mathcal{D}^{A_{n+1}}(\phi, K)$. Since $\bigcap_{n=1}^{\infty} \mathcal{D}^{A_n}(\phi, K)$ does not depend on the choice of $\{A_n\}$, we denote it by $\mathcal{D}^A(\phi, K)$. In case A is simultaneously closed and relatively open, i.e. isolated, this definition of $\mathcal{D}^A(\phi, K)$ coincides with the above one.

Now we formulate the Dirichlet principle as follows:

*Theorem* 1. *Suppose* $\mathcal{D}^A(\phi, K) \neq \emptyset$. *Then there exists a uniquely determined function* $h \in \mathcal{D}^A(\phi, K)$ *such that*

$$(*) \qquad (f - h, h)_{R-K} = 0$$

*for any* $f \in \mathcal{D}^A(\phi, K)$. *Furthermore,* h *is harmonic in* R - K.

*Proof:* (i) First we suppose that A is closed. Then, by the aid of Lemma 1, we can prove that $\phi_K^{\Omega_n}(P)$ tends to a harmonic function h locally uniformly and in norm on R - K as $n \to \infty$. It is easy to see that this function h satisfies the required properties.

(ii) Next we suppose that A is relatively open. By (i), there exists a uniquely determined function $h_n \in \mathcal{D}^{A_n}(\phi, K)$ which satisfies (*) for any $f \in \mathcal{D}^{A_n}(\phi, K)$. Since $\mathcal{D}^{A_n}(\phi, K) \supset \mathcal{D}^{A_{n+1}}(\phi, K)$, we see that $\{h_n\}$ form a Cauchy sequence in norm and $h_n$ tends to a harmonic function $h_0$ locally uniformly and in norm on $R - K$ as $n \to \infty$. It can be seen that $h_0$ satisfies the required properties.

We see that h has the smallest Dirichlet norm among the functions in $\mathcal{D}^A(\phi, K)$. We shall denote h(P) by $\phi_K^A(P)$. If $\phi \equiv 1$ on $\partial K$, then $\phi_K^A(P)$ will be denoted by $\omega^A(P; K)$.

The following properties follow from 1)- 4) and Theorem 1: Let $\phi$, $\phi_i$ (i = 1, 2) be given continuous functions on $\partial K$ such that $\mathcal{D}^A(\phi, K) \neq \emptyset$ and $\mathcal{D}^A(\phi_i, K) \neq \emptyset$ (i = 1, 2).

a) $(a_1\phi_1 + a_2\phi_2)_K^A = a_1(\phi_1)_K^A + a_2(\phi_2)_K^A$ for any real numbers $a_1$, $a_2$.

b) $0 < \omega^A(P; K) \leqq 1$.

c) If $\phi \geqq 0$ on $\partial K$, then $\phi_K^A \geqq 0$ on $R - K$.

d) If K, K' are regular compact sets such that $K \subset K'$, then $(\phi_K^A)_{K'}^A(P) = \phi_K^A(P)$ for $P \in R - K'$.

By a), b) and c), we have the maximum principle for $\phi_K^A$:

e) $|\phi_K^A(P)| \leqq \max_{\partial K} |\phi|$.

*Lemma* 2 (C. Constantinescu [1], p. 75). If $\omega^A(P; K)$ is not identically equal to one, then $\inf_{P \in R-K} \omega^A(P; K) = 0$.

By b), c), d) and this lemma, we have

*Lemma* 3. If $\omega^A(P; K)$ is not identically equal to one for some K, then so is it for all K.

*Definition.* Let A be a non-empty closed or relatively open subset of $\Delta$ such that $A \neq \Delta$. A is said to be *weakly negligible* if $\omega^A(P; K)$ is identically equal to one for some (hence any) regular compact set K. (C. Constantinescu [1] called this fact "halbschwach".)

By a), b), c) and Lemma 2, we have

*Lemma* 4. Suppose A is not weakly negligible. If u is a harmonic function on R such that $u = u_K^A$ on R - K for some K, then u is identically equal to zero.

## § 3. Function $N^A$

3.1 N-function for an end.

Let $\Omega$ be an end on R and let Q be a point in $\Omega$. By a discussion analogous to that in [3], we can prove that there exists a uniquely determined function $N(P, Q; \Omega)$ which has the following properties:

6) $N(P, Q; \Omega)$ is a positive harmonic function of P in $\Omega - \{Q\}$, vanishes on $\partial\Omega$ and has a logarithmic singularity with coefficient 1 at P = Q.

7) $N(P, Q; \Omega) = N(Q, P; \Omega)$.

8) If K is a regular compact set in $\Omega$ whose interior contains Q, then $(N(\cdot, Q; \Omega))_K^{\Omega}(P) = N(P, Q; \Omega)$ for $P \in \Omega - K$.

3.2 Function $N^A$.

First we suppose that A is closed. Then there is an approximation $\{\Omega_n\}$ of R such that $B(R - \Omega_n) \searrow A$ as $n \to \infty$.

We shall prove

*Lemma* 5. In order that the increasing limit of $N(P, Q; \Omega_n)$ be not identically equal to $\infty$, it is necessary and sufficient that A is not weakly negligible.

*Proof*: Let $n_o$ be any fixed integer. Let $D = \Omega_{n_o}$ and $C = \partial\Omega_{n_o}$. We write $\omega(P) = \omega^A(P; C)$, $\omega_n(P) = \omega(P; C, \Omega_n)$ and $N_n(P, Q) = N(P, Q; \Omega_n)$ $(n > n_o)$. By Green's formula we have

$$(1) \int_C N_n(P,Q)\frac{\partial\omega_n(P)}{\partial\nu_P}\, ds_P = \int_C \frac{\partial N_n(P,Q)}{\partial\nu_P}\, ds_P = \int_{\partial\Omega_n} \frac{\partial N_n(P,Q)}{\partial\nu_P}\, ds_P = 2\pi$$

$(n > n_o)$ for $Q \in D$, where $\nu$ is drawn inward with respect to D. As $n \to \infty$, $\omega_n$ tends to $\omega$ and $\frac{\partial \omega_n}{\partial \nu} \geqq 0$ decreases to $\frac{\partial \omega}{\partial \nu} \geqq 0$.

Necessity: Since $N_n(P, Q)$ converges to a function uniformly for $P \in C$, $\frac{\partial \omega}{\partial \nu}$ is not identically equal to zero by (1). Hence $\omega$ is not identically equal to one, i.e., A is not weakly negligible.

Sufficiency: Suppose A is not weakly negligible. Since $\|\omega_n\|_{\Omega_n} \geqq \|\omega\|_R > 0$, it follows from (1) that

$$\min_{P' \in C} N_n(P',Q)\|\omega\|_R^2 \leqq \min_{P' \in C} N_n(P',Q) \int_C \frac{\partial \omega_n}{\partial \nu} ds \leqq 2\pi$$

and hence $N_n(P, Q)$ does not tend to constant $\infty$ as $n \to \infty$.

The limit $\lim_{n \to \infty} N(P, Q; \Omega_n)$ does not depend on the choice of $\{\Omega_n\}$. Thus we shall denote it by $N^A(P, Q)$ if it is not identically equal to $\infty$. We note that $N^{B(R-\Omega_n)}(P, Q)$ tends to $N^A(P, Q)$ as $n \to \infty$.

Next we suppose that A is relatively open. Then there is a sequence $\{A_n\}$ of non-empty isolated subsets of $\Delta$ such that $A_n \nearrow A$ as $n \to \infty$. $N^{A_n}(P, Q)$ exists if A is not weakly negligible. Hence in this case, $\lim_n N^{A_n}(P, Q)$ exists as the limit of the decreasing sequence and denote the limit function by $N^A(P, Q)$.

The following properties follow from 5)-7) and the above definition:

f) $N^A(P, Q)$ is a positive harmonic function of P in $R - \{Q\}$ and has a logarithmic singularity with coefficient 1 at $P = Q$.

g) $N^A(P, Q) = N^A(Q, P)$.

h) If K is a regular compact set whose interior contains Q,

then $(N^A(\cdot, Q))_K^A(P) = N^A(P, Q)$ for $P \in R - K$.

## § 4. Ideal Boundary

4.1 Definition of ideal boundary points.

Let D be a non relatively compact domain on R with compact relative boundary (this may be empty). An extended real valued continuous function $M(P, Q)$ on $D \times D$ is called a *kernel* on D if it satisfies the following two conditions:

(i) For each $Q \in D$, $M(P, Q)$ is a positive harmonic function of P in $D - \{Q\}$ and has a logarithmic singularity at $P = Q$.

(ii) For each $P \in D$, there is a regular compact set K in R such that $\{P\} \cup \partial D \subset K - \partial K$ and $\sup_{Q \in D-K} M(P, Q) < \infty$.

Let $\mathcal{M}(D)$ be the set of all kernels on D and let M be an element of $\mathcal{M}(D)$. Let $\{Q_j\}$ be a sequence of points in D clustering nowhere in R. If $M(P, Q_j)$ converges to a harmonic function locally uniformly as $j \to \infty$, then $\{Q_j\}$ will be called a fundamental sequence with respect to M. If the limiting harmonic functions of two sequences $\{M(P, Q_j)\}$ and $\{M(P, Q'_j)\}$ are equal to each other, then we say that $\{Q_j\}$ and $\{Q'_j\}$ are equivalent with respect to M. This is an equivalence relation. We call an equivalence class an *M-boundary point* of D. We denote by $\Delta_M$ the set of all M-boundary points of D. Let Q be an M-boundary point. If $\{Q_j\}$ determines Q, then we set

$$M(P, Q) = \lim_{j \to \infty} M(P, Q_j);$$

This value does not depend on the choice of a fundamental sequence $\{Q_j\}$. We introduce a metric on $D \cup \Delta_M$ by

$$d(Q_1, Q_2) = \sup_{P \in K_0} \left| \frac{M(P, Q_1)}{1 + M(P, Q_1)} - \frac{M(P, Q_2)}{1 + M(P, Q_2)} \right|$$

for any $Q_1$, $Q_2 \in D \cup \Delta_M$, where $K_0$ is a closed disk in D. The topology induced by this metric on D is independent of the choice of $K_0$ and coincides with the original topology on D.

The metric space $D \cup \Delta_M$ has following properties:

(α) $\Delta_M$ is compact in $D \cup \Delta_M$.

(β) The function M(P, Q), for a fixed $P \in D$, is finite continuous as a function of Q in $D \cup \Delta_M - \{P\}$.

Let E be a non-empty isolated subset of B(D). Let Ω be an end on R such that $\Omega \subset D$ and $B(\Omega) = E$. Since the intersection of $\Delta_M$ and the closure of Ω in $D \cup \Delta_M$ does not depend on the choice of Ω, we denote it by $(\Delta_M)_E$. If A is the limit of some increasing sequence $\{E_n\}$ of isolated subsets of B(D), then we define $(\Delta_M)_A = \bigcup_n (\Delta_M)_{E_n}$.

4.2 Examples.

1. *Kuramochi boundary.* Let $K_0$ be a closed disk in R. Then the Kuramochi kernel $N(P, Q) \equiv N(P, Q; R - K_0)$ belongs to $\mathcal{M}(R - K_0)$. $\Delta_N$ is the Kuramochi boundary of R (Cf. [2] and [3]).

2. *Martin boundary.* Suppose R is a hyperbolic Riemann surface. Let G(P, Q) be the Green function on R and let $Q_0$ be a fixed point in R. Then the Martin kernel $K(P, Q) = G(P, Q)/G(Q_0, Q)$ $(Q \neq Q_0)$ belongs to $\mathcal{M}(R - \{Q_0\})$. $\Delta_K$ is the Martin boundary of R (Cf. [4]).

3. *Function N(P, Q; Ω).* By 6), 7), 8) and 5), we see that N(P, Q; Ω) belongs to $\mathcal{M}(\Omega)$.

4. *Function $N^A$.* By f), g), h) and e), we see that $N^A$ belongs to $\mathcal{M}(R)$.

5. *Function $K^A$.* Let $Q_0$ be a fixed point in R and set $K^A(P, Q) = N^A(P, Q)/N^A(Q_0, Q)$ $(Q \neq Q_0)$. Then, by f), g), h) and e), we see that $K^A$ belongs to $\mathcal{M}(R - \{Q_0\})$.

## § 5. Correspondence among boundaries

Suppose A is not weakly negligible in this section.

### 5.1 Correspondence between $\Delta_{K^A}$ and the Kuramochi boundary $\Delta_N$.

Suppose A is closed and let $\Omega$ be an end on R such that $B(\Omega) \subset \Delta - A$. As in the proof of Lemma 5, we see that

$$\int_{\partial\Omega} \frac{\partial N^A(P, Q)}{\partial \nu_P} ds_P = 2\pi \quad \text{or} \quad 0$$

according as $Q \in \Omega$ or $Q \in R - (\Omega \cup \partial\Omega)$. Hence we have

*Lemma* 6. Let $B = \Delta - A$.

(i) $N^A(P, Q) > 0$ for $Q \in (\Delta_{N^A})_B$.

(ii) If $Q_1$, $Q_2$ are two different points of $(\Delta_{N^A})_B$, then $N^A(P, Q_1)$ and $N^A(P, Q_2)$ are not proportional to each other.

We prove

*Theorem* 2. *Suppose* A *is closed and not weakly negligible. Let* $B = \Delta - A$. *Then there exists a homeomorphism of* $R \cup (\Delta_{N^A})_B$ *onto* $R \cup (\Delta_N)_B$ *which reduces to the identity on* R.

*Proof*: Let $\{\Omega_n\}$ be an approximation of R such that $B(R - \Omega_n) \searrow A$ and let $N_n(P, Q) = N(P, Q; \Omega_n)$. For $Q \in \Omega_n$ it holds that

$$(2) \qquad N^A(P, Q) - (N^A(\cdot, Q))^A_{\partial\Omega_n}(P) = N_n(P, Q) \quad \text{or} \quad 0$$

according as $P \in \Omega_n$ or $P \in R - \Omega_n$. As $Q \to Q' \in (\Delta_{N^A})_{B(\Omega_n)}$, $N^A(P, Q)$ tends to $N^A(P, Q')$ uniformly for $P \in \partial\Omega_n$. Hence, by e) we see that $N_n(P, Q)$ can be continuously extended over $(\Delta_{N^A})_{B(\Omega_n)}$ as a function of Q. We denote the extended function on $(\Delta_{N^A})_{B(\Omega_n)}$ by $N_n^*(P, Q)$.

Suppose $N_n^*(P, Q_1) \equiv N_n^*(P, Q_2)$ for $Q_1, Q_2 \in (\Delta_{N^A})_{B(\Omega_n)}$. Then, by (2), we have

$$N^A(P,Q_1) - N^A(P,Q_2) = (N^A(\cdot,Q_1) - N^A(\cdot,Q_2))^A_{\partial\Omega_n}(P)$$

for every point P in R. Hence we have $N^A(P, Q_1) \equiv N^A(P, Q_2)$ by Lemma 4. It follows that there exists a homeomorphism of $\Omega_n \cup (\Delta_{N^A})_{B(\Omega_n)}$ onto $\Omega_n \cup \Delta_{N_n}$ which reduces to the identity on $\Omega_n$. On the other hand, by a discussion analogous to that in Theorem 12 in [3], we can show that there exists a homeomorphism of $\Omega_n \cup \Delta_{N_n}$ onto $\Omega_n \cup (\Delta_N)_{B(\Omega_n)}$ which reduces to the identity on $\Omega_n$. Since n is arbitrary, we have the theorem.

*Theorem* 3. *Under the same assumption as in Theorem 2, there exists a homeomorphism of* $R \cup (\Delta_{K^A})_B$ *onto* $R \cup (\Delta_N)_B$ *which reduces to the identity on* R.

*Proof*: $K^A(P, Q)$ can be continuously extended over $(\Delta_{N^A})_B$ as a function of Q by (i) of Lemma 6. If

$$\lim_{Q\to Q_1} K^A(P, Q) \equiv \lim_{Q\to Q_2} K^A(P, Q)$$

for $Q_1, Q_2 \in (\Delta_{N^A})_B$, then $N^A(P, Q_2) \equiv aN^A(P, Q_1)$, where $a = N^A(Q_o, Q_2)/N^A(Q_o, Q_1) > 0$. Hence $N^A(P, Q_1) \equiv N^A(P, Q_2)$ by (ii) of Lemma 6. Thus we have the theorem by the aid of Theorem 2.

5.2 Correspondence between $\Delta_{K^A}$ and the Martin boundary $\Delta_K$.

*Theorem* 4. *Suppose* A *is relatively open and not weakly negligible. Then there exists a homeomorphism of* $R \cup (\Delta_{K^A})_A$ *onto*

$R \cup (\Delta_K)_A$ *which reduces to the identity on* R.

*Proof:* Let $\{\Omega_n\}$ be an approximation of R such that $Q_o \in \Omega_1$ and $B(\Omega_n) \nearrow A$ as $n \to \infty$. Let n be any fixed integer. Let g(P, Q) be the Green function on $\Omega_n$ and let

$$\tilde{K}(P, Q) = \frac{g(P, Q)}{N^A(Q_o, Q)}, \quad k(P, Q) = \frac{g(P, Q)}{g(Q_o, Q)} \quad (Q \neq Q_o).$$

Then it holds that

$$N^A(P, Q) - (N^A(\cdot, Q))^A_{\partial\Omega_n}(P) = g(P, Q) \quad \text{or} \quad 0$$

according as $P \in \Omega_n$ or $P \in R - \Omega_n$. Hence

$$(3) \quad K^A(P, Q) - (K^A(\cdot, Q))^A_{\partial\Omega_n}(P) = \tilde{K}(P, Q) \quad \text{or} \quad 0$$

according as $P \in \Omega_n$ or $P \in R - \Omega_n$. It follows that $\tilde{K}(P, Q)$ can be continuously extended over $(\Delta_{K^A})_{B(\Omega_n)}$ as a function of Q and the extended function $\tilde{K}^*(P, Q)$ is positive for every $Q \in (\Delta_{K^A})_{B(\Omega_n)}$. Hence $k(P, Q) = \tilde{K}(P, Q)/\tilde{K}(Q_o, Q)$ can be continuously extended over $(\Delta_{K^A})_{B(\Omega_n)}$ as a function of Q. Denote the extended function by $k^*(P, Q)$. If $k^*(P, Q_1) \equiv k^*(P, Q_2)$ for $Q_1, Q_2 \in (\Delta_{K^A})_{B(\Omega_n)}$, then $\tilde{K}^*(P, Q_2) \equiv a\tilde{K}^*(P, Q_1)$, where $a = \tilde{K}^*(Q_o, Q_2)/\tilde{K}^*(Q_o, Q_1) > 0$. Hence, by (3), we have

$$K^A(P, Q_2) - aK^A(P, Q_1) = (K^A(\cdot, Q_2) - aK^A(\cdot, Q_1))^A_{\partial\Omega_n}(P)$$

for every point P in R. Hence $K^A(P, Q_1) \equiv K^A(P, Q_2)$ by Lemma 4. Thus there exists a homeomorphism of $\Omega_n \cup (\Delta_{K^A})_{B(\Omega_n)}$ onto $\Omega_n \cup \Delta_k$

which reduces to the identity on $\Omega_n$. On the other hand, Parreau (see [4]) showed that there exists a homeomorphism of $\Omega_n \cup \Delta_k$ onto $\Omega_n \cup (\Delta_K)_{B(\Omega_n)}$ which reduces to the identity on $\Omega_n$. Therefore we have the theorem.

## *References*

[1] C. Constantinescu: Ideale Randkomponenten einer Riemannschen Fläche, Rev. Math. pures et appl., 4 (1959), 43-76.

[2] Z. Kuramochi: Potentials on Riemann surfaces, J. Fac. Sci. Hokkaido Univ. Ser. I, 16 (1962), 5-79.

[3] M. Ohtsuka: An elementary introduction of Kuramochi boundary, J. Sci. Hiroshima Univ. Ser. A-I Math., 28 (1964), 271-299.

[4] M. Parreau: Sur les moyennes des fonctions harmoniques et analytiques et la classification des surfaces de Riemann, Ann. Inst. Fourier, 3 (1952), 103-197.

Department of Mathematics,
Faculty of Science,
Okayama University

# IV. ON BEURLING'S AND FATOU'S THEOREMS

Zenjiro KURAMOCHI 1)

## Introduction

There are two typical theorems, Fatou's and Beurling's on the boundary behavior of analytic functions in $|z| < 1$. Many extensions of these theorems to the case of Riemann surfaces have been made; see [1], [2], [3], [4]. However, different methods have been used to prove the above two theorems. In this paper we shall give a unified method to prove that, for some analytic mapping of a Riemann surface with N-Martin (= Kuramochi) boundary or K-Martin (= Martin) boundary into another Riemann surface, fine limits exist on the boundary except on a set of capacity zero or of harmonic measure zero.

The results were already published in [7] and [8]. We shall try to present a more rigorous version.

## § 1. Capacitary potential and harmonic measure

Let R be an open Riemann surface. A subset of R will be said to have a *piecewise analytic (relative) boundary* if the relative boundary consists of an enumerable number of analytic arcs which cluster nowhere in R. A continuous function in an open set G will be called *piecewise smooth* if it is continuously differentiable in an open subset $G' \subset G$ such that $G - G'$ locally consists of a finite number of points and open analytic arcs. Let G be an open subset of R with piecewise analytic boundary $\partial G$ and $\phi$ be a continuous function on $\partial G$. Denote by $\mathfrak{D}_{\phi}^{G}$ the family of piecewise smooth functions f in G with boundary values $\phi$ on $\partial G$ and with finite Dirichlet integral

---

1) This paper was thoroughly revised by F-Y. Maeda and M. Ohtsuka.

$$\|f\|^2 = \iint\left\{\left(\frac{\partial f}{\partial x}\right)^2 + \left(\frac{\partial f}{\partial y}\right)^2\right\}dxdy < \infty.$$

Suppose $\mathfrak{D}_\phi^G$ is not empty. Take an exhaustion $\{R_n\}$ of R with compact analytic boundaries $\partial R_n$, and let $h_n$ be the harmonic function in $G \cap R_n$ which has the boundary values $\phi$ on the closure of $\partial G \cap R_n$ and whose normal derivative vanishes on the rest of the boundary. One can show that $h_n$ converges to a function in $\mathfrak{D}_\phi^G$ locally uniformly and in Dirichlet norm; see Theorem 5 of [11]. Let us denote it by $H_\phi^G$. This has the smallest Dirichlet norm among the functions of $\mathfrak{D}_\phi^G$. We note that $H_{\phi'}^{G'} = H_\phi^G$ for any $G' \subset G$ with piecewise analytic boundary, where $\phi' = \phi$ on $\partial G' \cap \partial G$ and $= H_\phi^G$ on $\partial G' \cap G$; see Theorem 3 of [11].

We shall use the notions related to extremal length and Fuglede's lemma; see [11] for them. We have

*Lemma* 1. Let G be an open set with piecewise analytic boundary and $\phi$ be a continuous boundary function on $\partial G$ such that $\mathfrak{D}_\phi^G$ is not empty. Then $\int_c \partial H_\phi^G/\partial\nu ds = 0$ for a.e. c which is the piecewise analytic relative boundary $\partial G_c$ of some domain $G_c$ such that $G_c \cup \partial G_c \subset G$ and $\partial G_c$ is also the boundary of $R - G_c \cup \partial G_c$.

*Proof.* Let $\{R_n\}$ be an exhaustion of R and $h_n$ be the harmonic function constructed above. Then $h_n$ converges to $H_\phi^G$ is Dirichlet norm. By Green's formula

$$\int_{c \cap R_n} \frac{\partial h_n}{\partial \nu} ds = 0$$

for any $c = \partial G_c$. Extend $h_n$ by zero to $G - R_n$ and denote the function in G by $h_n$ again. By Fuglede's lemma, taking a subsequence if necessary,

$$\left|\int_c \frac{\partial H_\phi^G}{\partial \nu} ds\right| = \left|\int_c \frac{\partial H_\phi^G}{\partial \nu} ds - \int_c \frac{\partial h_n}{\partial \nu} ds\right| \leqq \int_c |\text{grad}\,(h_n - H_\phi^G)|ds \to 0$$

as $n \to \infty$ for a.e. c.

Consider the case that there is given a continuous boundary function $\phi$ on $\partial G$ which takes 0 or 1. If $\phi = 1$ on a closed set $A \subset \partial G$ and $= 0$ on $\partial G - A$ and if $\mathfrak{D}_{\phi}^{G}$ is not empty, then $H_{\phi}^{G}$ will be denoted by $\omega(A, z, G)$. Let $\{F_k\}$ be a decreasing sequence of closed sets, each of which is contained in G and has a piecewise analytic boundary, such that $\bigcap_k F_k = \emptyset$. If $\omega_k(z) = \omega(\partial F_k, z, G - F_k)$ exists for some k, then $\omega_k$ converges locally uniformly and in Dirichlet norm to a harmonic function in G. We shall denote it by $\omega(\{F_k\}, z, G)$. This is equal to $H_f^{\{(R-G)\cup F_k\}}$ of [11], where f is a piecewise smooth Dirichlet finite function in G which is equal to 0 on $\partial G$ and to 1 on $F_1$. In the following lemma we assume that $\omega(\{F_k\}, z, G)$ exists and denote it by $\omega(z)$. By Theorem 10 of [11], $\sup \omega = 1$ unless $\omega \equiv 0$.

*Lemma* 2. a) Assume $\omega \not\equiv 0$ and set $G_{t_1,t_2} = \{z \in G;\ 0 < t_1 < \omega(z) < t_2 < 1\}$. Then $(\omega(z) - t_1)/(t_2 - t_1)$ equals $\omega(C_{t_2}, z, G_{t_1,t_2})$, where $C_t$ is the level curve $\{z \in G;\ \omega(z) = t\}$.

b) $\int_{C_t} \partial\omega/\partial\nu ds = \|\omega\|^2$ for almost every t, $0 < t < 1$.

*Proof.* a) The extremal distance with respect to $G_{t_1,t_2} - F_k$ between $\partial F_k \cap G_{t_1,t_2}$ and a compact set K in $G_{t_1,t_2}$ which is bounded by a closed analytic curve, tends to $\infty$ as $k \to \infty$ on account of Theorem 8 of [11]. By making use of Theorem 9 of [11] we see that the harmonic function with minimum Dirichlet integral in $G_{t_1,t_2}$, which is equal to $\omega$ on $\partial G_{t_1,t_2}$, coincides with $\omega$.

b) Choose $t_1$, $t_2$ such that $0 < t_1 < t_2 < 1$, denote $\omega(C_{t_2}, z, G_{t_1,t_2})$ by $\omega_1(z)$ simply. We can show as in Proposition (iii) of [12] that

$$\int_{\{\omega_1=t\}} \frac{\partial\omega}{\partial\nu}\, ds = \|\omega_1\|^2$$

for almost all t, $0 < t < 1$. Substituting $\omega_1 = (\omega - t_1)/(t_2 - t_1)$ we obtain

$$\frac{1}{t_2-t_1}\int_{\{\omega=t\}} \frac{\partial\omega}{\partial\nu}\, ds = \frac{\|\omega\|^2_{G_{t_1,t_2}}}{(t_2-t_1)^2} \qquad \text{for almost all } t,\ t_1 < t < t_2.$$

By letting $t_1 \to 0$ and $t_2 \to 1$ we have $\int_{C_t} \partial\omega/\partial\nu\, ds = \|\omega\|^2$ for almost all t, $0 < t < 1$.

Any enumerable family C of curves in G for which $\int_C \partial\omega/\partial\nu\, ds = \|\omega\|^2$ is called *complete* with respect to $\omega$.

*Lemma* 3. Let G be an open set with piecewise analytic boundary and $\phi$ be a continuous bounded function on $\partial G$ such that $\mathfrak{D}^G_\phi$ is not empty. Suppose $\omega(\{F_k\}, z, G) \not\equiv 0$ exists and denote it by $\omega(z)$. Then

$$\int_{C_t} H^G_\phi \frac{\partial\omega}{\partial\nu}\, ds$$

is constant for almost all t, $0 < t < 1$, where $C_t$ is a level curve for $\omega(z)$.

*Proof.* Choose $t_1$, $t_2$ such that $0 < t_1 < t_2 < 1$ and both $C_{t_1}$ and $C_{t_2}$ are complete. Let $\omega_n$ be the harmonic function in $G_{t_1,t_2} \cap R_n$ such that $\omega_n = \omega$ on the closure of $\partial G_{t_1,t_2} \cap R_n$ and $\partial\omega_n/\partial\nu = 0$ on the rest of the boundary. By Lemma 2, a), $\omega_n$ tends to $\omega$ locally uniformly and in Dirichlet norm. Extend $\omega_n$ to $G_{t_1,t_2} - R_n$ by 0. As $n \to \infty$

$$(t_2-t_1)\int_{C_{t_i}} \frac{\partial\omega_n}{\partial\nu}\,ds = \|\omega_n\|^2 + \|\omega\|^2_{G_{t_1,t_2}} = (t_2-t_1)\int_{C_{t_i}} \frac{\partial\omega}{\partial\nu}\,ds \quad (i=1,2).$$

Given $\varepsilon > 0$, choose $n_o$ such that $\int_{C_{t_i}-R_{n_o}} (\partial\omega/\partial\nu)ds < \varepsilon/3$ and $\left|\int_{C_{t_i}} (\partial\omega_n/\partial\nu - \partial\omega/\partial\nu)ds\right| < \varepsilon/3$ if $n \geqq n_o$. Since $\partial\omega_n/\partial\nu$ converges to $\partial\omega/\partial\nu$ uniformly on $C_{t_i}\cap R_{n_o}$, there exists $n_1 \geqq n_o$ such that

$$\int_{C_{t_i}\cap R_{n_o}} \left|\frac{\partial\omega_n}{\partial\nu} - \frac{\partial\omega}{\partial\nu}\right| ds < \frac{\varepsilon}{3} \qquad \text{for any } n \geqq n_1.$$

It holds that

$$\left|\int_{C_{t_i}-R_{n_o}} \frac{\partial\omega_n}{\partial\nu}\,ds\right|$$

$$= \left|\int_{C_{t_i}} \left(\frac{\partial\omega_n}{\partial\nu} - \frac{\partial\omega}{\partial\nu}\right)ds + \int_{C_{t_i}\cap R_{n_o}} \left(\frac{\partial\omega}{\partial\nu} - \frac{\partial\omega_n}{\partial\nu}\right)ds + \int_{C_{t_i}-R_{n_o}} \frac{\partial\omega}{\partial\nu}\,ds\right| < \varepsilon$$

if $n \geqq n_1$. Let $\sup|\phi| = M$. As $n \to \infty$, $\int_{C_{t_i}} H^G_\phi(\partial\omega_n/\partial\nu)ds$ tends to $\int_{C_{t_i}} H^G_\phi(\partial\omega/\partial\nu)ds$, because

$$\left|\int_{C_{t_i}} H^G_\phi \frac{\partial\omega}{\partial\nu}\,ds - \int_{C_{t_i}} H^G_\phi \frac{\partial\omega_n}{\partial\nu}\,ds\right| \leqq \int_{C_{t_i}\cap R_{n_o}} H^G_\phi\left|\frac{\partial\omega}{\partial\nu} - \frac{\partial\omega_n}{\partial\nu}\right|ds + 2\varepsilon M < 3\varepsilon M$$

if $n \geqq n_1$. Let $h_n$ be the harmonic function in $G_{t_1,t_2}\cap R_n$ such that

$h_n = H_\phi^G$ on the closure of $\partial G_{t_1,t_2} \cap R_n$ and $\partial h_n/\partial\nu = 0$ on the rest of the boundary. It tends to $H_\phi^G$ in Dirichlet norm. By Green's formula

$$\int_{C_{t_2}} H_\phi^G \frac{\partial\omega_n}{\partial\nu}\, ds - \int_{C_{t_1}} H_\phi^G \frac{\partial\omega_n}{\partial\nu}\, ds = (t_2-t_1)\int_{C_t\cap R_n} \frac{\partial h_n}{\partial\nu}\, ds,$$

where t is any number between $t_1$ and $t_2$. By Fuglede's lemma there is a subsequence $\{n_k\}$ such that $\int_{C_t\cap R_{n_k}} (\partial h_{n_k}/\partial\nu)ds$ tends to $\int_{C_t} \partial H_\phi^G/\partial\nu ds$ for a.e. t, $t_1 < t < t_2$ (cf. the proof of Lemma 1). By Lemma 1 $\int_{C_t} (\partial H_\phi^G/\partial\nu)ds = 0$ for a.e. t, $0 < t < 1$. It follows that

$$\int_{C_{t_1}} H_\phi^G \frac{\partial\omega}{\partial\nu}\, ds = \int_{C_{t_2}} H_\phi^G \frac{\partial\omega}{\partial\nu}\, ds.$$

This shows that $\int_{C_t} H_\phi^G(\partial\omega/\partial\nu)ds$ is constant for almost all t, $0 < t < 1$.

We shall consider a special example of $\omega(\{F_k\}, z, G)$. Let F be a closed set, contained in G, with piecewise analytic boundary. When $\omega(\{F-R_n\}, z, G)$ exists, it will be denoted by $\omega(B(F), z, G)$ and called the *capacitary potential* of the ideal boundary of F with respect to G. We see easily

$$\omega(B(\bigcup_{i=1}^{j} F_i), z, G) \leqq \sum_{i=1}^{j} \omega(B(F_i), z, G)$$

for closed subsets $F_1, \ldots, F_j$ of G with piecewise analytic boundaries, provided that each $\omega(B(F_i), z, G)$ is well-defined.

Finally we shall define a harmonic measure. Let G be any open set with $\partial G \neq \emptyset$ and A be any closed subset of $\partial G$. Denote by $w(A, z, G)$ the *harmonic measure* of A with respect to G. For any closed set F contained in G, $w(\partial(F-R_n), z, G-(F-R_n))$ decreases as $n \to \infty$. The limit will be denoted by $w(B(F), z, G)$ and called the *harmonic measure* of the ideal boundary of F with respect to G. We have

$$w(B(\bigcup_{i=1}^{j} F_i), z, G) \leqq \sum_{i=1}^{j} w(B(F_i), z, G).$$

## § 2. Fine neighborhoods of Kuramochi and Martin boundary points

A compact set bounded by a closed analytic curve will be called an analytic compact set. Let $N(z, p)$ be a function N or N-Green's function outside an analytic compact set $K_o$ in R, and define a metric on $R - R_1$ by

$$\sup_{z \in R_1 - K_o} \left| \frac{N(z,p_1)}{1+N(z,p_1)} - \frac{N(z,p_2)}{1+N(z,p_2)} \right| ,$$

where $R_1$ is a relatively compact open set containing $K_o$. Similarly define a metric on $R - R_1$ by

$$\sup_{z \in R_1} \left| \frac{K(z,p_1)}{1+K(z,p_1)} - \frac{K(z,p_2)}{1+K(z,p_2)} \right| ,$$

where $K(z, p)$ is the ratio of Green's functions $G(z, p)$ and $G(p, z_o)$. These metrics induce topologies on $R - R_1$ compatible with the original topology. Any metric on R which is compatible with the original topology and is equivalent to the above metric on $R - R_1$ will be called an N-Martin (= Kuramochi) metric or K-Martin (= Martin) metric. By completing R with respect to these metrics, compactifications of R are obtained.

The boundaries of R in these compactifications will be denoted by $B^N$, and $B^K$.[2] $B^{\alpha}$ ($\alpha$ = N or K) is decomposed into $B_0^{\alpha}$ and $B_1^{\alpha}$, where $B_1^{\alpha}$ is the set of $\alpha$-minimal points. It is known that $B_0^{\alpha}$ is an $F_{\sigma}$ set.[3]

Let E be a closed set in $B^{\alpha}$. Set $E_k = \{z \in R;\ \mathrm{dist}\,(z, E) \leqq 1/k\}$, where the distance is defined with respect to an $\alpha$-Martin metric. For each k we can find a closed set $E_k'$ in R with analytic boundary such that $E_{k+1} \subset E_k' \subset E_k - \partial E_k$; cf. [10], p. 289. Let F be a closed set in R with piecewise analytic boundary and G be an open set in R with piecewise analytic boundary containing F. If $\omega(\{E_k' \cap F\}, z, G)$ is well defined, it will be denoted by $\omega(E \cap B(F), z, G)$. Similarly $w(E \cap B(F), z, G)$ is defined by $\lim_{k\to\infty} w(\partial(E_k' \cap F), z, G-(E_k' \cap F))$.

We can show that $\omega(E, z) \equiv \omega(E \cap B(R-R_1), z, R-K_0)$ defines a capacity (in the sense of Choquet) on $B^N$ for each $z \in R - K_0$,[3] where $K_0$ is an analytic compact set and $R_1$ is an open set which contains $K_0$ and whose closure is an analytic compact set. Throughout the rest of this paper, we fix such $K_0$ and $R_1$, unless otherwise stated, and we shall use the notation $\omega(E, z)$ introduced above. A Borel set A on $B^N$ will be said to be of capacity zero if $\omega(E, z) = 0$ for any closed subset E of A, and otherwise of positive capacity.

The set function $w(E, z) = w(E \cap B(R), z, R)$ defines a measure on $B^K$ for each $z \in R$. This is called the harmonic measure.

It is known that $B_0^N$ is of capacity zero and $B_0^K$ is of harmonic measure zero.[3]

---

2) These are so-called Kuramochi and Martin boundaries respectively.
3) See [3], [6] and [10] for these facts.

Let U be a positive full-superharmonic function[3)] on $R - K_o$. Given a compact set $K \subset R - K_o$ with piecewise analytic boundary, let ${}_KU$ be the full-superharmonic function in $R - K_o$ satisfying ${}_KU = U$ on K and $= 0$ on $\partial K_o$ and having the following property: For any compact set K' with piecewise analytic boundary such that $K \cup K_o \subset \partial K' - \partial K'$, it holds that ${}_KU = H_\phi^{R-K'}$, where $\phi = {}_KU$ on $\partial K'$. If F is a closed set in $R - K_o$ with piecewise analytic boundary, then we define ${}_FU = \lim_{n\to\infty} {}_{F\cap(R_n \cup \partial R_n)}U$ for an exhaustion $\{R_n\}$ of R. We can show that this limit exists and equals $U_{\tilde{F}}$ defined at p. 164 of [3]; cf. § 5 of [10] too. We also infer from Theorem 11 of [10] that, in case U is piecewise smooth and Dirichlet finite in $R - K_o$, ${}_FU$ is equal to $H_\phi^{R-F-K_o}$ in $R-F-K_o$, where $\phi$ is equal to U on $\partial F$ and to 0 on $\partial K_o$.

Next let U be a positive superharmonic function on R. For any closed set F in R with piecewise analytic boundary, let $U_F$ be the least positive superharmonic function in R such that $U_F \geqq U$ on F.

Let G be any open subset of R and let $p \in B_1^\alpha$. If there is an open subset G' of $G - R_1$ (G resp.) with piecewise analytic boundary such that $N(z, p) \not\equiv {}_{R-R_1-G'}N(z, p)$ ($K(z, p) \not\equiv K(z, p)_{R-G'}$, resp.), then we say that R - G is *N-thin* (*K-thin*, resp.) at p and express this fact by $p \overset{N}{\not\in} G$ ($p \overset{K}{\not\in} G$, resp.). We call G an *N-fine* (a *K-fine*, resp.) *neighborhood* in R of p. We refer to [3] for properties of thin sets.

We now prove

*Lemma* 4. Let E be a closed subset of $B^\alpha$ ($\alpha$ = N or K). Let G be an open subset of R with piecewise analytic boundary; it is required that $G \cup \partial G \subset R-R_1 \cup \partial R_1$ if $\alpha$ = N. If there exists a closed subset F of G with piecewise analytic boundary such that

$\omega(z) = \omega(E \cap B(F), z, G)$ exists and $\not\equiv 0$, in case $\alpha = N$;

$w(z) = w(E \cap B(F), z, G) \not\equiv 0$, in case $\alpha = K$,

then there exists at least one point $p \in E \cap B_1^\alpha$ such that $p \notin G$.

*Proof.* We may assume that G is a domain. First let $\alpha = N$. $\omega(E, z)$ is a positive full-superharmonic function and $\omega(E, z) \geqq \omega(z)$. Let $U(z) = {}_{R-R_1-G}\omega(E, z)$ in G. As remarked before, $U(z) = H^G_{\omega(E, z)}$ in G. Hence by Lemma 3 $\int_{C_t} U(\partial\omega/\partial\nu)ds \equiv a$ (const.) for almost all t, $0 < t < 1$, where $C_t$ is a level curve of $\omega$. Choose $t_1$, $0 < t_1 < 1$, such that $C_{t_1}$ is complete (with respect to $\omega$) and $\int_{C_{t_1}} U(\partial\omega/\partial\nu)ds = a$. Since $\partial\omega/\partial\nu > 0$ and $U(z) \leqq \omega(E, z) < 1$ on $C_{t_1}$,

$$\int_{C_{t_1}} U \frac{\partial\omega}{\partial\nu} ds < \int_{C_{t_1}} \frac{\partial\omega}{\partial\nu} ds = \|\omega\|^2.$$

We can find $t_2$ close to 1 such that $C_{t_2}$ is complete,

$$\int_{C_{t_2}} U \frac{\partial\omega}{\partial\nu} ds = a \qquad \text{and} \qquad \int_{C_{t_1}} U \frac{\partial\omega}{\partial\nu} ds < t_2\|\omega\|^2.$$

Then

$$\int_{C_{t_2}} U \frac{\partial\omega}{\partial\nu} ds = a = \int_{C_{t_1}} U \frac{\partial\omega}{\partial\nu} ds < t_2\|\omega\|^2 = \int_{C_{t_2}} \omega \frac{\partial\omega}{\partial\nu} ds.$$

Hence there exists a point $z_o$ on $C_{t_2}$ such that $U(z_o) < \omega(z_o)$. Since $\omega(z) \leqq \omega(E, z)$, we have ${}_{R-R_1-G}\omega(E, z) < \omega(E, z)$ in G. Now $\omega(E, z)$ is represented uniquely by the potential $\int N(z, p)d\mu(p)$

of a positive measure $\mu$ on $B^N$ such that $\mu(B^N) = \mu(E \cap B_1^N)$.[3)]
Using Theorem 15 of [10], we obtain

$$\int N(z, p)d\mu(p) = \omega(E, z) > {}_{R-R_1-G}\omega(E, z) = \int\{{}_{R-R_1-G}N(z, p)\}d\mu(p).$$

Hence there exists at least one point $p \in E \cap B_1^N$ such that $p \overset{N}{\notin} G$.

Next we consider the case $\alpha = K$. Obviously, $w(E, z) - w(E, z)_{R-G} \geqq w(z)$ on $G$. Since $w(z) > 0$, we have $w(E, z) > w(E, z)_{R-G}$. By an argument similar to the above, replacing $N(p, z)$ by $K(p, z)$, we conclude that there exists at least one point $p \in E \cap B_1^K$ such that $p \overset{K}{\notin} G$.

## § 3. Function-theoretic separative metrics

Suppose a metric $d$ is given on $R$ which is compatible with the original topology of $R$. Denote by $B$ the boundary obtained by the completion with respect to $d$. For any two sets $S_1$ and $S_2$ in $R \cup B$, let $d(S_1, S_2)$ be the distance between $S_1$ and $S_2$.

If the metric $d$ satisfies the following condition D or B,[4)] it is called *H.D.* or *H.B. separative* respectively:

*Condition D.* For any pair $(F, G)$ of a closed set $F$ in $R$ and an open set $G$ in $R$, both having piecewise analytic boundaries, such that $G \cup \partial G \subset R-R_1 \cup \partial R_1$, $F \subset G$ and $d(F, R-G) > 0$, it holds that

$$\lim_{\varepsilon \to 0} \omega(B(F \cap F_{1-\varepsilon}), z, R-K_0) = 0,$$

where $F_{1-\varepsilon} = \{z \in G;\ {}_{R-R_1-G}\omega(B(F), z, R-K_0) \geqq 1 - \varepsilon\}$.

---

4) These are slightly different from the conditions originally given by the author; cf. [7]. In fact the present conditions are stronger than the original ones but not essentially different.

*Condition B.* For any pair (F, G) of a closed set F in R and an open set G in R, both having piecewise analytic boundaries, such that $F \subset G$ and $d(F, R-G) > 0$, it holds that

$$\lim_{\varepsilon \to 0} w(B(F \cap F_{1-\varepsilon}), z, R) = 0,$$

where $F_{1-\varepsilon} = \{z \in G;\ w(\partial G, z, G) \geqq 1 - \varepsilon\}$.

Remark. Condition D does not depend on the choice of $K_0$ and $R_1$. In fact, independence on $R_1$ is obvious. Now suppose $K_0 \subset K_0' \subset R_1$. Let $\omega(z) = \omega(B(F), z, R-K_0)$, $\omega'(z) = \omega(B(F), z, R-K_0')$, $F_{1-\varepsilon}' = \{z \in G;\ {}_{R-R_1-G}\omega'(z) \geqq 1-\varepsilon\}$, $\omega_\varepsilon(z) = \omega(B(F \cap F_{1-\varepsilon}), z, R-K_0)$ and $\omega_\varepsilon'(z) = \omega(B(F \cap F_{1-\varepsilon}'), z, R-K_0')$. Since $F_{1-\varepsilon} \supset F_{1-\varepsilon}'$, $\omega_\varepsilon(z) \geqq \omega_\varepsilon'(z)$. Hence $\lim_{\varepsilon \to 0}\omega_\varepsilon(z) = 0$ implies $\lim_{\varepsilon \to 0}\omega_\varepsilon'(z) = 0$. To show the converse, consider $M = \sup_{z \in \partial K_0'}\omega(z)$ and $M_\varepsilon = \sup_{z \in \partial K_0'}\omega_\varepsilon(z)$. Then $M, M_\varepsilon < 1$. Since $(1 - M)\omega'(z) + M \geqq \omega(z)$ on $R-K_0'$, $(1-M)_{R-R_1-G}\omega'(z) + M \geqq {}_{R-R_1-G}\omega(z)$. It follows that $F_{1-\varepsilon} \subset F_{1-\varepsilon'}'$, where $\varepsilon' = \varepsilon/(1-M)$. Hence $\lim_{\varepsilon \to 0}\omega_\varepsilon'(z) = 0$ implies $\lim_{\varepsilon \to 0}\omega(B(F \cap F_{1-\varepsilon}), z, R-K_0') = 0$. We shall show that this implies $\lim_{\varepsilon \to 0}\omega_\varepsilon(z) = 0$. Let $\tilde{\omega}(z) = \lim_{\varepsilon \to 0}\omega_\varepsilon(z)$ and let $\tilde{M} = \sup_{z \in \partial K_0'}\omega_0(z)$. Then $\tilde{M} = \lim_{\varepsilon \to 0}M_\varepsilon$. From

$$(1-M_\varepsilon)\omega(B(F \cap F_{1-\varepsilon}), z, R-K_0') + M_\varepsilon \geqq \omega_\varepsilon(z) \qquad \text{on } R-K_0',$$

it follows that $\tilde{M} \geqq \tilde{\omega}(z)$ on $R-K_0'$. By maximum principle, we conclude that $\tilde{\omega} \equiv 0$.

If R is a Riemann surface with null boundary, then $\omega(B(F \cap F_{1-\varepsilon}), z, R-K_0) \equiv 0$ for any $\varepsilon$, so that any metric on R is H.D. separative. Let us show that any N-Martin metric $d_N$ is H.D.

separative and any K-Martin metric $d_K$ is H.B. separative.[5)]

1) N-Martin metric. Let (F, G) be a pair in Condition D for $d_N$. We must show that $\tilde{\omega}(z) = \lim_{\varepsilon\to 0}\omega(B(F\cap F_{1-\varepsilon}), z, R-K_0) = 0$. Suppose the contrary. Then we have $0 < \|\tilde{\omega}\| < \infty$. Since $\tilde{\omega}$ and ${}_{R-R_1-G}\omega$ are both full-superharmonic, they are expressed uniquely as potentials $\int N(z, p)d\mu(p)$ and $\int N(z, p)d\nu(p)$ respectively, where the support of $\mu$ ($\nu$, resp.) is contained in the closure of F (R-G, resp.) in $R\cup B^N$ and $\mu$ and $\nu$ are canonical.[6)] Since $d_N(F, R-G) > 0$, $\mu$ and $\nu$ are different measures, and hence $\tilde{\omega} \neq {}_{R-R_1-G}\tilde{\omega}$. Let $\omega' = \tilde{\omega} - {}_{R-R_1-G}\tilde{\omega}$, $M = \sup \omega'(z)$ and $E_M = \{z\in G;\ \omega'(z) \geqq M/2\}$. Then $M > 0$ and we have

$$\omega'(z) \leqq \frac{M}{2}\,\omega(\partial(F_{1-\varepsilon} - (E_M - \partial E_M)), z) + \omega(\partial(F\cap F_{1-\varepsilon}\cap E_M), z)$$

in G, where $\omega(\partial A, z)$ means $\omega(\partial A, z, R-K_0-A)$ for a closed set A with piecewise analytic boundary such that $A\subset R-K_0$. It then follows that $\lim_{\varepsilon\to 0}\omega(\partial(F\cap F_{1-\varepsilon}\cap E_M), z) \not\equiv 0$, since, otherwise, $\omega'(z) \leqq M/2$, which is impossible. Since the function min $(2\omega'/M, 1)$ has finite Dirichlet integral and is equal to 1 on $E_M$, to 0 on $\partial G$, $\omega_\varepsilon(z) = \omega(\partial(F\cap F_{1-\varepsilon}\cap E_M)$ $z, G-(F\cap F_{1-\varepsilon}\cap E_M))$ exists and $\|\omega_\varepsilon\| \leqq \|\min(2\omega'/M, 1)\| < \infty$. It follows that $\hat{\omega}(z) = \lim_{\varepsilon\to 0}\omega_\varepsilon(z)$ exists and the convergence is in Dirichlet norm. On the other hand, by Dirichlet principle

$$\|\omega(\partial(F\cap F_{1-\varepsilon}\cap E_M), z)\| \leqq \|\omega_\varepsilon\|.$$

---

5) More generally, it was shown by H. Tanaka that a metric d is H.B. separative if the completion of R with respect to d is a resolutive compactification ([3]) of R. In particular, any N-Martin metric $d_N$ is also H.B. separative.

6) A measure $\mu$ on $R\cup B^\alpha$ is called canonical if $\mu(B^\alpha) = \mu(B_1^\alpha)$.

Since $\lim_{\varepsilon \to 0} \omega(\partial(F \cap F_{1-\varepsilon} \cap E_M), z) \not\equiv 0$, we conclude that $\hat{\omega}(z) \not\equiv 0$. Obviously, $\hat{\omega} \leqq {}_{R-R_1-G}\,\omega$. By Lemma 3,

$$\int_{\{\hat{\omega}=t\}} ({}_{R-R_1-G}\omega) \frac{\partial \hat{\omega}}{\partial \nu} ds = a \text{ (const.) for almost all } t,\ 0 < t < 1.$$

Set $\alpha = \|\hat{\omega}\|^2 - a$. Since ${}_{R-R_1-G}\omega < 1$ on $G$, $\alpha > 0$. It holds that

$$\|\hat{\omega}\|^2 - \alpha = \int_{\{\hat{\omega}=t\}} ({}_{R-R_1-G}\omega) \frac{\partial \hat{\omega}}{\partial \nu} ds \geqq \int_{\{\hat{\omega}=t\}} \hat{\omega} \frac{\partial \hat{\omega}}{\partial \nu} ds = t\|\hat{\omega}\|^2$$

for almost all $t$, $0 < t < 1$. This absurd if $t$ is sufficiently close to 1. Thus we conclude that $\tilde{\omega} \equiv 0$.

2) K-Martin metric. Let $(F, G)$ be a pair in Condition B for $d_K$. Let $A_1$ be the intersection of the closure of $F$ in $R \cup B^K$ with $B^K$ and let $A_2$ be the closure of $R - G$ in $R \cup B^K$. Then $A_1 \cap A_2 = \emptyset$. Let $v(z) = 1$ on $R - G$ and $= w(\partial G, z, G)$ on $G$. Then $v$ is a positive superharmonic function on $R$ and

$$v_{A_1}(z) \geqq (1 - \varepsilon) w(B(F \cap F_{1-\varepsilon}), z, R),$$

where $v_{A_1}(z) = \lim_{n\to\infty} v_{F-R_n}(z)$ for an exhaustion $\{R_n\}$ of $R$. The difference $v - v_{A_1}$ is also non-negative superharmonic on $R$. Thus we have unique expressions $v(z) = \int K(z, p) d\mu(p)$, $v_{A_1}(z) = \int K(z, p) d\mu_1(p)$ and $v(z) - v_{A_1}(z) = \int K(z, p) d\nu(p)$ with canonical measures[6] $\mu$, $\mu_1$ and $\nu$ such that the support of $\mu$ ($\mu_1$, resp.) is contained in $A_2$ ($A_1$, resp.). Since $\nu + \mu_1 = \mu$, it follows that $\mu_1 = 0$, i.e., $v_{A_1} = 0$. Hence $w(B(F \cap F_{1-\varepsilon}), z, R) = 0$ for any $\varepsilon > 0$.

*Lemma* 5. If $d$ is an H.D. separative metric on $R$, then, for any pair $(F, G)$ in Condition D, there exists a sequence $\{V_n\}$ of relatively open subsets of $F$ with piecewise analytic boundaries such that

$\omega(B(F-V_n), z, R-K_o)$ tends to zero locally uniformly and in Dirichlet norm ($n \to \infty$) and such that each $\omega(\partial V_n, z, G-V_n-\partial V_n)$ exists (unless $V_n = \emptyset$).

*Proof.* Let $\omega(z) = \omega(B(F), z, R-K_o)$ and let

$$\tilde{V}_n = \{z \in G;\ {}_{R-R_1-G}\,\omega(z) < 1 - \tfrac{1}{n}\},$$

$$\Omega_n = \{z \quad G;\ \omega(z) > 1 - \tfrac{1}{2n}\}$$

and

$$V_n = \tilde{V}_n \cap \Omega_n \cap F.$$

Then each $V_n$ is a relatively open subset of F with piecewise analytic boundary. Since $G - \tilde{V}_n = F_{1-(1/n)}$, Condition D implies $\lim_{n\to\infty} \omega(B(F-\tilde{V}_n), z, R-K_o) = 0$. It is easy to see that this convergence is also in Dirichlet norm. On the other hand,

$$\omega(B(F-V_n), z, R-K_o) \leqq \omega(B(F-\tilde{V}_n), z, R-K_o) + \omega(B(F-\Omega_n), z, R-K_o).$$

We shall show that $\omega(B(F-\Omega_n), z, R-K_o) = 0$ for each n. Then we have $\omega(B(F-V_n), z, R-K_o) = \omega(B(F-\tilde{V}_n), z, R-K_o) \to 0$ ($n \to \infty$) locally uniformly and in Dirichlet norm.

To show that $\omega(B(F-\Omega_n), z, R-K_o) = 0$, we use Theorem 8 of [11]. By this theorem we see that the extremal distance $\lambda_k^{(n)}$ between $\partial K_o$ and $F-\Omega_n-R_k$ increases to $\infty$ as $k \to \infty$ for each n. It is known that $\lambda_k^{(n)} = \|\omega(\partial(F-\Omega_n-R_k, z, R-(F-\Omega_n-R_k))\|^{-2}$. Hence $\omega(B(F-\Omega_n), z, R-K_o) = 0$ for every n.

Finally consider the function $\min\,(2n(\omega - {}_{R-R_1-G}\,\omega), 1)$ on G. This function is Dirichlet finite on G, equal to 1 on $\partial V_n$ and to 0 on $\partial G$. Hence $\omega(\partial V_n, z, G-V_n-\partial V_n)$ exists.

## § 4. Fine cluster sets

Let $\zeta = f(z)$ be an analytic mapping of a Riemann surface R with positive boundary into another Riemann surface $\underline{R}$ which may have a positive or null boundary. By means of f(z), R is regarded as a covering surface of $\underline{R}$. We consider the α-Martin boundary $B^\alpha$ (α = N or K) of R and a metric d compatible with the original topology on $\underline{R}$. Let $\underline{B}$ be the boundary of $\underline{R}$ obtained by the completion with respect to d.

For $p \in B_1^\alpha$ we put

$$\overset{\alpha}{M}(f(p)) = \bigcap_\tau \overline{f(G_\tau)},$$

where $G_\tau$ runs over all α-fine neighborhoods in R of p and the closure $\overline{f(G_\tau)}$ is taken in $\underline{R} \cup \underline{B}$. We shall call $\overset{\alpha}{M}(f(p))$ the *fine cluster set* of f at p. If $\underline{R} \cup \underline{B}$ is compact then $\overset{\alpha}{M}(f(p))$ is non-empty for any $p \in B_1^\alpha$ and consists of one point or a continuum in $\underline{R} \cup \underline{B}$; see [3], pp. 146 and 221.

*Lemma* 6. (i) Let G be an open set with piecewise analytic boundary in R($G \cup \partial G \subset R - R_1 \cup \partial R_1$ if α = N). Then $\{p \in B_1^\alpha;\ p \not\overset{\alpha}{\in} G\}$ is a $G_\delta$ subset of $B^\alpha$.

(ii) $S = \{p \in B_1^\alpha,\ \text{diam}\ \overset{\alpha}{M}(f(p)) > 0\}$ is a $G_{\delta\sigma}$ subset of $B^\alpha$.

*Proof*. We shall prove our lemma in case α = N. The proof for α = K is quite analogous. For any fixed $z \in R$, N(z, p) is a continuous function of p on $B^N$ and ${}_{R-R_1-G}N(z, p)$ is lower semi-continuous on $B^N$. Therefore $\{p \in B^N;\ N(z, p) = {}_{R-R_1-G}N(z, p)\}$ is a $G_\delta$ set. Since $\{p \in B_1^N;\ p \not\overset{N}{\in} G\}$ is the intersection of the above set with $B_1^N$ and since $B_1^N$ is a $G_\delta$ set, we have assertion (i). Next choose a sequence $\{\zeta_i\}$ of points which are everywhere dense in $\underline{R}$. For each i, there exists a sequence $\{\underline{G}_{i,n}\}_n$ of open sets with piecewise analytic boundaries in $\underline{R}$ such that $D(\zeta_i, 1/2n) \subset \underline{G}_{i,n} \subset D(\zeta_i, 1/n)$, $n = 1, 2, \ldots$, where $D(p, r) = \{\zeta \in \underline{R};\ d(\zeta, p) < r\}$ for $r > 0$. Then we can show that

$$S = \bigcup_{n=1}^{\infty} \bigcap_{i=1}^{\infty} \{p \in B_1^N;\ p \overset{N}{\notin} f^{-1}(\underline{G}_{i,n})\}.$$

Therefore, (i) implies that S is a $G_{\delta\sigma}$ set.

## § 5. Extensions of Beurling's and Fatou's theorems

Let $\zeta = f(z)$ be an analytic mapping of R into $\underline{R}$. If f(z) satisfies the following conditions then the covering surface over $\underline{R}$ defined by f(z) is called *almost finitely sheeted.*

1) For $\zeta \in \underline{R}$, let $n(\zeta)$ be the number of times that $\zeta$ is covered by the covering surface. If we take a sufficiently large compact set $\underline{K}$, then $n(\zeta) \leq M < \infty$ in $\underline{R} - \underline{K}$.

2) For any point p of $\underline{R}$, there exists a set $\Delta_p \subset \underline{R}$ mapped onto a compact disk by a local parameter at p such that the part of the covering surface lying over $\Delta_p$ has a finite total area measured with respect to the local parameter.

We shall prove

*Theorem* 1. (Extension of beurling's theorem) *Let* R *be a Riemann surface with positive boundary and consider the N-Martin boundary* $B^N$ *of* R. *Let* $\zeta = f(z)$ *be an analytic mapping of* R *into another Riemann surface* $\underline{R}$ *which may be with positive or null boundary, and suppose that the corresponding covering surface is almost finitely sheeted. Consider an* H.D. *separative metric on* $\underline{R}$ *and define the boundary* $\underline{B}$ *by the completion with respect to the metric. Assume that* $\underline{R} \cup \underline{B}$ *is compact. Then* $S = \{p \in B_1^N;\ \text{diam}\ \overset{N}{M}(f(p)) > 0\}$ *is a* $G_{\delta\sigma}$ *set of capacity zero.*

*Proof.* By Lemma 6, S is a $G_{\delta\sigma}$ set. Since $\underline{R} \cup \underline{B}$ is compact, for each positive integer n, we can choose a finite number of open sets $D_{n,1}, \ldots, D_{n,m_n}$ in $\underline{R}$ with piecewise analytic boundaries such that each $D_{n,j}$ has diameter less than 2/n and any disk D(p, 1/(6n)) is contained in some $D_{n,j}$. Then we can show that

$$S = \bigcup_{n=1}^{\infty} \bigcap_{j=1}^{m_n} \{p \in B_1^N;\ p \overset{N}{\notin} f^{-1}(D_{n,j})\}.$$

Assume that S is of positive capacity. Then there exists $n_o$ and a compact set E in the Borel set $\bigcap_{j=1}^{m_{n_o}} \{p \in B_1^N;\ p \overset{N}{\notin} f^{-1}(D_{n_o,j})\}$ such that E has a positive capacity, i.e., $\omega(E, z) > 0$. Let $\{\underline{R}_n\}$ be an exhaustion of $\underline{R}$ and set $\underline{A}_n = \underline{R} - \underline{R}_n$. Set $A_n = f^{-1}(\underline{A}_n)$ and define $\omega(E \cap B', z)$ by $\lim_{n\to\infty} \omega(E \cap B(A_n), z, R-K_o)$. We distinguish two cases.

Case 1. $\omega(E \cap B', z) > 0$. Since $f(R_1 \cup \partial R_1)$ is compact, there exists a number $n_1 \geqq n_o$ such that $f(R_1 \cup \partial R_1) \cap \underline{A}_{n_1} = \emptyset$ and $n(\zeta) \leqq M < \infty$ in $\underline{A}_{n_1}$. Since $\underline{R} \cup \underline{B}$ is compact, we can find a finite number of points $p_1, \ldots, p_k$ in $\underline{R}$ and positive numbers r, r' such that $r < r' \leqq 1/(6n_o)$ and

$$\underline{A}_{2n_1} \subset \bigcup_{i=1}^{k} D(p_i, r) \subset \bigcup_{i=1}^{k} D(p_i, r') \subset \underline{A}_{n_1}.$$

For each i, we can choose a closed set $\underline{F}_i$ and an open set $\underline{G}_i$, both with piecewise analytic boundaries, such that

$$D(p_i, r) \subset \underline{F}_i \subset \underline{G}_i \subset D(p_i, r')$$

and $d(\underline{F}_i, \underline{R} - \underline{G}_i) > 0$.

We now make the following convention: For any closed set F with piecewise analytic boundary contained in $R - K_0$, let $\omega(E \cap B(F), z) \equiv \omega(E \cap B(F), z, R-K_0)$ and $\omega(E \cap B(F) \cap B', z) = \lim_{n\to\infty} \omega(E \cap B(F \cap A_n), z)$.

Since $\bigcup_{i=1}^{k} f^{-1}(\underline{F}_i) \supset A_{2n_1}$, we have

$$\omega(E \cap B', z) \leqq \sum_{i=1}^{k} \omega(E \cap B(f^{-1}(\underline{F}_i)) \cap B', z).$$

Hence, there exists $i_0$ such that

(1) $$\omega(E \cap B(f^{-1}(\underline{F}_{i_0})) \cap B', z) > 0.$$

We shall write $\underline{F}$ for $\underline{F}_{i_0}$, $\underline{G}$ for $\underline{G}_{i_0}$ and $G$ for $f^{-1}(\underline{G})$. We may choose $\underline{K}_0 = \underline{R}_{n_1} \cup \partial \underline{R}_{n_1}$ in Condition D (cf. the remark after Condition D). Then the pair $(\underline{F}, \underline{G})$ has the properties stated in Condition D. Hence, by Lemma 5, there exists a sequence $\{V_q\}$ of relatively open subsets of $\underline{F}$ with piecewise analytic boundaries such that

(2) $$\omega(B(\underline{F} - V_q), \zeta, \underline{R}-\underline{K}_0) \to 0$$

as $q \to \infty$, locally uniformly and in Dirichlet norm, and such that each $\omega(\partial V_q, \zeta, \underline{G}-V_q-\partial V_q)$ exists. Put $U_{n,q}(z) = \omega(\partial((\underline{F}-V_q) \cap \underline{A}_n), f(z), \underline{R}-\underline{K}_0-(\underline{F}-V_q) \cap \underline{A}_n)$ on $f^{-1}(\underline{R}-K_0-(\underline{F}-V_q) \cap \underline{A}_n)$, $U_{n,q}(z) = 1$ on $f^{-1}((\underline{F}-V_q) \cap \underline{A}_n)$ and $U_{n,q}(z) = 0$ on $f^{-1}(\underline{K}_0)$ for $n > n_1$. Since $n(\zeta) \leqq M < \infty$ on $\underline{R} - \underline{K}_0$,

$$\|U_{n,q}(z)\| = M\|\omega(\partial((\underline{F}-V_q) \cap \underline{A}_n), \zeta, \underline{R}-\underline{K}_0-(\underline{F}-V_q) \cap \underline{A}_n)\| < \infty.$$

Hence, by Dirichlet principle and by letting $n \to \infty$, we see that $\omega(E \cap B(f^{-1}(\underline{F}-V_q)) \cap B', z)$ exists and

$$\|\omega(E \cap B(f^{-1}(\underline{F}-V_q)) \cap B', z)\| \leqq M\|\omega(B(\underline{F}-V_q), \zeta, \underline{R}-\underline{K}_0)\|.$$

Hence, by (2),

(3) $$\lim_{q\to\infty} \| \omega(E \cap B(f^{-1}(\underline{F}-V_q)) \cap B', z) \| = 0.$$

Next, set $u_q(z) = \omega(\partial V_q, f(z), G-V_q-\partial V_q)$ on $f^{-1}(\underline{G}-V_q)$, $u_q(z) = 1$ on $f^{-1}(V_q)$ and $u_q(z) = 0$ on R-G. As above, we have $\|u_q(z)\| \leqq M \|\omega(\partial V_q, \zeta, \underline{G}-V_q \cup \partial V_q)\| < \infty$. Hence $\omega(E \cap B(f^{-1}(V_q \cup \partial V_q))$, z, G) exists. On the other hand, it holds by (1) that

$$0 < \omega(E \cap B(f^{-1}(\underline{F})) \cap B', z)$$

$$\leqq \omega(E \cap B(f^{-1}(V_q \cup \partial V_q)) \cap B', z) + \omega(E \cap B(f^{-1}(\underline{F}-V_q)) \cap B', z).$$

Hence, by (3), there exists $q_o$ such that

$$\omega(E \cap B(f^{-1}(V_{q_o} \cup \partial V_{q_o})) \cap B', z) > 0.$$

Set $F = f^{-1}(V_{q_o} \cup \partial V_{q_o})$. By Dirichlet principle, we have

$$0 < \|\omega(E \cap B(F) \cap B', z)\| \leqq \|\omega(E \cap B(F), z, G)\|.$$

Hence $\omega(E \cap B(F), z, G) \not\equiv 0$.

Case 2. $\omega(E \cap B', z) = 0$. In this case, there exists $n' > n_1$ such that $\underline{R}_{n'} \supset f(R_2 \cup \partial R_2)$ and $\omega(E \cap B(R-(A_{n'}-\partial A_{n'})-R_2), z) > 0$, where $R_2$ is a relatively compact open set in R with analytic boundary such that $R_2 \supset R_1 \cup \partial R_1$. Since $\underline{R}_{n'}$ is relatively compact, we can choose a finite number of points $p_1, \ldots, p_k$ and positive numbers r, r' such that $r < r' \leqq 1/(6n_o)$, each $D(p_i, r')$ is relatively compact, the closure of each $D(p_i, r')$ is disjoint from $f(R_1 \cup \partial R_1)$ and $(\underline{R}_{n'} \cup \partial \underline{R}_{n'}) - f(R_2) \subset \bigcup_{i=1}^{k} D(p_i, r)$. For each i, choose $\underline{F}_i$ and $\underline{G}_i$ as in Case 1. Then there exists $i_o$ such that $\omega(E \cap B(f^{-1}(\underline{F}_{i_o})), z) > 0$. Set $\underline{F} = \underline{F}_{i_o}$,

$\underline{G} = \underline{G}_{1_0}$, $F = f^{-1}(\underline{F})$ and $G = f^{-1}(\underline{G})$. Since $\underline{F}$ is compact, there exists a continuously differentiable function $U(\zeta)$ on $\underline{R}$ which has a finite Dirichlet integral, which is equal to 1 on $\underline{F}$ and whose support is contained in $\underline{G}$. Put $u(z) = U(f(z))$ in $G$ and $u(z) = 0$ in $R-G$. Since $f$ is almost finitely sheeted, condition 2) stated just before Theorem 1 implies that $\|u(z)\| < \infty$. Since $u = 1$ on $F$ and $= 0$ on $\partial G$, it follows that $\omega(E \cap B(F), z, G)$ exists. By Dirichlet principle,

$$0 < \|\omega(E \cap B(F), z)\| \leqq \|\omega(E \cap B(F), z, G)\|.$$

Hence $\omega(E \cap B(F), z, G) \not\equiv 0$.

Thus in any case we find a closed set $F$ and an open set $G$ in $R$, both having piecewise analytic boundaries, such that $F \subset G$, $G \cup \partial G \subset R-R_1 \cup \partial R_1$, $f(G) \subset D(p, 1/(6n_0))$ for some $p \in \underline{R}$ and $\omega(E \cap B(F), z, G)$ exists and $\not\equiv 0$. By the choice of $\{D_{n,j}\}$, there exists $j_0$ such that $f(G) \subset D_{n_0,j_0}$, i.e., $G \subset f^{-1}(D_{n_0,j_0})$. On the other hand, by Lemma 4, there exists at least one point $p_0 \in E \cap B_1^N$ such that $p_0 \overset{N}{\notin} G$. Therefore $p_0 \overset{N}{\notin} f^{-1}(D_{n_0,j_0})$. This contradicts the relation $p_0 \in E \subset \bigcap_j \{p \in B_1^N;\ p \overset{N}{\notin} f^{-1}(D_{n_0,j})\}$. Thus $S$ is a set of capacity zero.

In a similar way, we prove the following

***Theorem* 2, a).** (Extension of Fatou's theorem) *Let $R$ be a Riemann surface with positive boundary and consider the K-Martin boundary $B^K$ of $R$. Let $\zeta = f(z)$ be an analytic mapping of $R$ into another Riemann surface $\underline{R}$ with positive boundary and consider an H.B. separative metric on $\underline{R}$ such that the completion $\underline{R} \cup \underline{B}$ with respect to this metric is compact. Then $S = \{p \in B_1^N;\ \operatorname{diam} \overset{K}{M}(f(p)) > 0\}$ is a $G_{\delta\sigma}$ set of harmonic measure zero*

*Proof.* By Lemma 6, $S$ is a $G_{\delta\sigma}$ set. We choose open sets $\{D_{n,j}\}$ as in the previous proof. Assume that $S$ is of positive harmonic measure. Then, as in the previous proof, there exist $n_o$ and a compact set E in $\bigcap_{j=1}^{m_n}\{p \in B_1^K;\ p \notin f^{-1}(D_{n_o,j})\}$ such that $w(E, z) > 0$. Let $A_n$ be as in the previous proof and write $w(E \cap B(F), z) \equiv w(E \cap B(F), z, R)$, $w(E \cap B', z) \equiv \lim_{n\to\infty} w(E \cap B(A_n), z)$ and $w(E \cap B(F) \cap B', z) \equiv \lim_{n\to\infty} w(E \cap B(F \cap A_n), z)$ for a closed set F in R with piecewise analytic boundary.

Case 1. $w(E \cap B', z) > 0$. Let $0 < r < r' \leqq 1/(6n_o)$ and choose a finite number of points $p_1, \ldots, p_k$ in $\underline{R}$ such that $\underline{R} = \bigcup_{i=1}^{k} D(p_i, r)$. For each i, choose $\underline{F}_i$ and $\underline{G}_i$ as in the previous proof. Then

$$w(E \cap B', z) \leqq \sum_{i=1}^{k} w(E \cap B(f^{-1}(\underline{F}_i)) \cap B', z),$$

and hence there exists $i_o$ such that

$$w(E \cap B(f^{-1}(\underline{F}_{i_o})) \cap B', z) > 0.$$

Set $\underline{F} = \underline{F}_{i_o}$, $\underline{G} = \underline{G}_{i_o}$, $F = f^{-1}(\underline{F})$ and $G = f^{-1}(\underline{G})$. For $\varepsilon > 0$, set

$$\delta_{1-\varepsilon} = \{z \in G;\ w(E \cap B(F) \cap B', z)_{R-G} \geqq 1 - \varepsilon\}$$

and

$$\underline{F}_{1-\varepsilon} = \{\zeta \in \underline{G};\ w(\partial \underline{G}, \zeta, \underline{G}) \geqq 1 - \varepsilon\}.$$

Since $w(E \cap B(F) \cap B', z)_{R-G} \leqq w(\partial \underline{G}, f(z), \underline{G})$ on G, we have $\delta_{1-\varepsilon} \subset f^{-1}(\underline{F}_{1-\varepsilon})$. It follows that

$$w(E \cap B(F \cap \delta_{1-\varepsilon}) \cap B', z) \leqq w(B(\underline{F} \cap \underline{F}_{1-\varepsilon}), f(z)).$$

Since the metric d is H.B. separative and the pair $(\underline{F}, \underline{G})$ has the properties in Condition B, $\lim_{\varepsilon\to o} w(B(\underline{F} \cap \underline{F}_{1-\varepsilon}), \zeta) = 0$. Hence $\lim_{\varepsilon\to o} w(E \cap B(F \cap \delta_{1-\varepsilon}) \cap B', z) = 0$. Therefore there exists $\varepsilon_o > 0$

such that $w(E\cap B(F\cap\delta')\cap B', z) > 0$, where $\delta' = \{z\in G;\ w(E\cap B(F)\cap B', z)_{R-G} \leqq 1 - \varepsilon_0\}$. Consider the set $\Omega = \{z\in G;\ w(E\cap B(F)\cap B', z) \geqq 1 - \varepsilon_0/2\}$. Then we can easily show that $w(E\cap B(F-(\Omega-\partial\Omega))\cap B', z) \leqq 1 - \varepsilon_0/2$ in R. It follows that $w(E\cap B(F-(\Omega-\partial\Omega))\cap B', z) \equiv 0$. Hence

$$w(E\cap B(F\cap\delta'\cap\Omega)\cap B', z) = w(E\cap B(F\cap\delta')\cap B', z) > 0.$$

Therefore $\delta'\cap\Omega = \emptyset$, which means $w(E\cap B(F)\cap B', z) \not\equiv w(E\cap B(F)\cap B', z)_{R-G}$. Hence $w(E\cap B(F)\cap B', z, G) = w(E\cap B(F)\cap B', z) - w(E\cap B(F)\cap B', z)_{R-G} \not\equiv 0$, and hence $w(E\cap B(F), z, G) \not\equiv 0$.

Case 2. Replacing $\omega$ by w in Case 2 of the previous proof, we can choose a closed set $\underline{F}$ and an open set $\underline{G}$ in $\underline{R}$, both having piecewise analytic boundaries, such that $\underline{F}\subset\underline{G}$, $\underline{G}$ is relatively compact and is contained in a set $D(p, 1/(6n_0))$ for some $p\in\underline{R}$ and such that $w(E\cap B(f^{-1}(\underline{F})), z) > 0$. Set $F = f^{-1}(\underline{F})$ and $G = f^{-1}(\underline{G})$. Obviously, $w(E\cap B(F), z) \leqq w(\partial\underline{F}, f(z), \underline{R}-\underline{F})$ on R - F. Since $\underline{R}$ has positive boundary, $w(\partial\underline{F}, \zeta, \underline{R}-\underline{F}) \not\equiv 1$. Since $\partial\underline{G}$ is compact in $\underline{R} - \underline{F}$, there exists a component $\underline{C}$ of $\partial\underline{G}$ such that

$$\eta = \sup_{\zeta\in\underline{C}} w(\partial\underline{F}, \zeta, \underline{R}-\underline{F}) < 1.$$

It follows that $w(E\cap B(F), z)_{R-G} \leqq \eta < 1$ on a component of G. Hence

$$w(E\cap B(F), z, G) = w(E\cap B(F), z) - w(E\cap B(F), z)_{R-G} \not\equiv 0.$$

Now, replacing N by K in the previous proof, we complete the present proof in the same way.

In case R is the unit disk, the original Fatou's theorem follows from the above theorem and the fact that if $\overset{K}{M}(f(p))$ consists of one point $\underline{q}\in\underline{R}\cup\underline{B}$, then there exists a path terminating at p and having

$\underline{q}$ as an asymptotic value; see [3] for this fact.

Let R be a Riemann surface with positive boundary, let $(R^{\infty}, \phi)$ be the universal covering surface of R and map it onto D: $|Z| < 1$ by $\psi$. Then $z(Z) = \phi(\psi^{-1}(Z))$ is an analytic mapping of D onto R. Given an analytic mapping f(z) of R into a Riemann surface $\underline{R}$ with null boundary, if $\zeta = f(z(Z))$ has an angular limit a.e. on $|Z| = 1$, then we call f(z) a function of F-type. For instance, if f(z) is of bounded type, then we can show that it is of F-type by using the original Fatou's theorem (see [5], [9], [3]).

Finally we prove

*Theorem* 2, b). (Extension of Fatou's theorem) *Let* R *be a Riemann surface with positive boundary and consider the K-Martin boundary* $B^K$ *of* R. *Let* $\underline{R}$ *be a Riemann surface with null boundary and let* $\zeta = f(z)$ *be an analytic mapping of* R *into* $\underline{R}$. *If* f(z) *is of F-type, then* $S = \{p \in B_1^K;\ \text{diam}\ \overset{K}{M}(f(p)) > 0\}$ *is a* $G_{\delta\sigma}$ *set of harmonic measure zero.*

*Proof.* Choose a metric d on $\underline{R}$ which is compatible with the topology of $\underline{R}$ and with respect to which the completion $\underline{R} \cup \underline{B}$ is compact (e.g., the K-Martin metric). Supposing the present theorem is not true, we find a closed set $E \subset B_1^K$ having the same properties as in the proof of the previous theorem; in particular, $w(E, z) > 0$. Then we find a closed set $\underline{F}$ and an open set $\underline{G}$ in $\underline{R}$, both having piecewise analytic boundaries, such that $\underline{F} \subset \underline{G}$, $d(\underline{F}, \underline{R}-\underline{G}) > 0$, diam $\underline{G} \leqq 1/(3n_0)$ and $w(E \cap B(F), z, R) > 0$, where $F = f^{-1}(\underline{F})$. It is enough to show that $w(E \cap B(F), z, G) \not\equiv 0$, where $G = f^{-1}(\underline{G})$.

Let $z = z(Z)$ be the analytic mapping of D: $|Z| < 1$ onto R, which is considered above. Let A be the subset of C: $|Z| = 1$ where f(z(Z)) has angular limits contained in $\underline{F}$. We shall show that $w(E \cap B(F), z, R)$ tends to 0 radially at a.e. point of C - A.

We may assume that $m(C-A) > 0$, where m is the linear measure on C. By assumption $\zeta = f(z(Z))$ has an angular limit a.e. on C. Hence, given $\varepsilon > 0$, we can find a closed set $\underline{F}_\varepsilon$ in $\underline{R}$ such that $d(\underline{F}_\varepsilon, \underline{F}) > 0$, $m(A'_\varepsilon) > 0$ and $m(C-A-A'_\varepsilon) < \varepsilon$ where $A'_\varepsilon$ is the set of points on C at which $f(z(Z))$ has angular limit lying on $\underline{F}_\varepsilon$.

For $\delta > 0$ and $e^{i\theta} \in C$, let $G(\delta, e^{i\theta}) = \{Z \in D;\ |\arg(1-e^{-i\theta}Z)| < \pi/4,\ 1-\delta < |Z| < 1\}$. If B is a subset of C, then we shall define $G(\delta, B) = \bigcup_{e^{i\theta}\in B} G(\delta, e^{i\theta})$. Consider the function

$$\phi_\delta(e^{i\theta}) = d(f(z(G(\delta, e^{i\theta}))), \underline{F})$$

defined on $A'_\varepsilon$. Then $\lim_{\delta\to 0}\phi_\delta(e^{i\theta})$ exists and is positive for each $e^{i\theta} \in A'_\varepsilon$. Using Egoroff's theorem, we can find a non-empty closed set $A_\varepsilon \subset A'_\varepsilon$ and $\delta > 0$ such that $m(C-A-A_\varepsilon) < \varepsilon$ and $G(\delta, A_\varepsilon)$ is disjoint from $z^{-1}(F)$.

It is easy to see that $G(\delta, A_\varepsilon)$ is bounded by rectifiable curves. Therefore, by F. and M. Riesz's theorem, a Borel set on $\partial G(\delta, A_\varepsilon)$ is of harmonic measure zero with respect to $G(\delta, A_\varepsilon)$ if and only if it has linear measure zero. For a sufficiently small $\delta > 0$, each component $G_0$ of $G(\delta, A_\varepsilon)$ is a Jordan domain. Let $Z_0 = e^{i\theta} \in A_\varepsilon$ be on the boundary of $G_0$ and let $c = \{Z;\ \arg Z = \theta,\ 1 - \frac{\delta}{2} < |Z| < 1\}$. Consider the harmonic measure W of an arc on $\partial G_0$ with one end point at $Z_0$ and the other in D. Since the angular domain $G(\delta, e^{i\theta})$ is contained in $G_0$, we can show that $0 < \inf_c W \leq \sup_c W < 1$. It follows that if we map $G_0$ onto the unit disk conformally, then the image of c is included in an angular domain. Thus we see that the harmonic measure $w_0$ of $\partial G(\delta, A_\varepsilon) \cap D$ in $G(\delta, A_\varepsilon)$ tends to zero radially at a.e. point of $A_\varepsilon$. Let $w_n(z) = w(\partial(F-R_n), z, R-(F-R_n))$ for an exhaustion $\{R_n\}$ of R. It is easy to see that $w_n(z(Z))$ is equal to the harmonic measure of $\partial(z^{-1}(F-R_n))$ in $D - z^{-1}(F-R_n)$. On the other hand, the extension of $w_0$ by 1 is a positive superharmonic function in D and is equal to 1 on $z^{-1}(F-R_n)$. Hence $w_o(Z) \geqq w_n(z(Z))$. Since $w(B(F), z, R) = \lim_{n\to\infty} w_n(z)$, we have $w_o(Z) \geqq w(B(F), z(Z), R) \geqq w(E \cap B(F), z(Z), R)$ in D. Therefore $w(E \cap B(F), z(Z), R)$ tends to 0 radially at a.e. point of $A_\varepsilon$. Since $\varepsilon$ is arbitrary, the same is true at a.e. point of C - A.

Since $w(E \cap B(F), z(Z), R) > 0$, we infer that $m(A) > 0$ and, by an argument similar to the above using Egoroff's theorem, we can choose $\delta_o > 0$ and a closed set $A' \subset A$ such that $m(A') > 0$, $w(E \cap B(F), z(Z), R) > \delta_o$ in $G(\delta_o, A')$, $d(f(z(Z)), \underline{F}) < d(\underline{F}, \underline{R} - \underline{G})/2$ in $G(\delta_o, A')$ and a Green function $g(z(Z), z_o)$ tends to zero uniformly as $Z \in G(\delta_o, A')$ tends to A'. Let $w^*(Z)$ be the harmonic measure of $\partial G(\delta_o, A') \cap D$ in $G(\delta_o, A')$. As was shown for $w_o$ above, we see that $w^*(Z)$ tends to 0 radially at a.e. point of A'.

Now let $u_n(z)$ be the harmonic function in $G \cap R_n$ which is equal to 0 on $\partial R_n \cap G$ and to $w(E \cap B(F), z, R)$ on $\partial G \cap R_n$. As $n \to \infty$, $u_n(z)$ increases to $w(E \cap B(F), z, R)_{R-G}$. We shall show that $u_n(z(Z)) \leqq w^*(Z)$ in $G_n = z^{-1}(G \cap R_n) \cap G(\delta_o, A')$. Since $d(f(z(Z)), \underline{F}) < d(\underline{F}, \underline{R} - \underline{G})/2$ in $G(\delta_o, A')$, $G(\delta_o, A') \subset z^{-1}(G)$. Hence $G_n = z^{-1}(R_n) \cap G(\delta_o, A')$. Since $\inf_{R_n} g(z, z_o) > 0$, $\partial G_n \cap A' = \emptyset$. Thus $u_n(z(Z)) = 0 \leqq w^*(Z)$ on $z^{-1}(\partial R_n) \cap G(\delta_o, A')$ and $u_n(z(Z)) \leqq 1 = w^*(Z)$ on $z^{-1}(R_n) \cap \partial G(\delta_o, A')$. Therefore $u_n(z(Z)) \leqq w^*(Z)$ in $G_n$. It then follows that $w(E \cap B(F), z(Z), R)_{R-G} \leqq w^*(Z)$ in $G(\delta_o, A')$. Since $w^*(Z)$ tends to 0 radially at a.e. point of A', so does $w(E \cap B(F), z(Z), R)_{R-G}$. On the other hand, $w(E \cap B(F), z(Z), R) > \delta_o$ in $G(\delta_o, A')$. Hence $w(E \cap B(F), z, R)_{R-G} \not\equiv w(E \quad B(F), z, R)$, or equivalently, $w(E \cap B(F), z, G) \not\equiv 0$.

*References*

[1] C. Constantinescu and A. Cornea: Über das Verhalten der analytischen Abbildungen Riemannscher Flächen auf dem idealen Rand von Martin, Nagoya Math. J., 17 (1960), 1-87.

[2] C. Constantinescu and A. Cornea: Le théorème de Beurling et la frontière idéale de Kuramochi, C. R. Acad. Sci. Paris, 254 (1962), 1732-1734.

[3] C. Constantinescu and A. Cornea: Ideale Ränder Riemannscher Flächen, Springer (1963).

[4] J. L. Doob: Conformally invariant cluster value theory, Illinois J. Math., 5 (1961), 521-547.

[5] Z. Kuramochi: On covering surfaces, Osaka Math. J., 5 (1953), 155-201.

[6] Z. Kuramochi: Potentials on Riemann surfaces, J. Fac. Sci. Hokkaido Univ. Ser. I, 16 (1962), 5-79.

[7] Z. Kuramochi: On the behaviour of analytic functions on the ideal boundary, I-IV, Proc. Japan Acad., 38 (1962), 150-155, 188-203.

[8] Z. Kuramochi: Correction to the paper "On the behaviour of analytic functions", ibid., 39 (1963), 27-32.

[9] M. Ohtsuka: Reading of the paper "On covering surfaces" by Z. Kuramochi, Mimeographed, Univ. of Kansas, 1959.

[10] M. Ohtsuka: An elementary introduction of Kuramochi boundary, J. Sci. Hiroshima Univ. Ser. A-I Math., 28 (1964), 271-299.

[11] M. Ohtsuka: Dirichlet principle on Riemann surfaces, J. Analyse Math., 19 (1967), 295-311.

[12] M. Ohtsuka: On Kuramochi's paper "Potentials on Riemann surfaces", these Proceedings, No. 5.

Department of Mathematics,
Faculty of Science,
Hokkaido University

# V. ON KURAMOCHI'S PAPER "POTENTIALS ON RIEMANN SURFACES"[1)]

Makoto OHTSUKA

## Introduction

In 1956 Z. Kuramochi [2] defined a new boundary for any open Riemann surface R. It is now called Kuramochi boundary. Rigorous treatments of this boundary are found in [1] and [4]. It shares some properties with inner points of R. For instance, the values of every SHS function[2)] are defined on the Kuramochi boundary in [1]. This was also done by Kuramochi [3] but his discussions were not quite clear.

The purpose of the present paper is to make Kuramochi's definition of the values on the boundary rigorous, following his discussions in [3]. We take a closed "disk" $K_o$ in R and a closed set F which is disjoint from $K_o$. There exists a harmonic function $\omega_F$ in $R - K_o - F$ which takes 0 on $\partial K_o$ and 1 on $\partial F$ and has the smallest Dirichlet integral. If $\int_{\partial F} \partial\omega_F/\partial\nu \, ds$ is equal to the Dirichlet integral of $\omega_F$, then $\partial F$ is called complete. As customarily, a function N(P, Q) is defined for every Kuramochi boundary point Q. If Q is minimal, the value at Q of any SHS function V is defined by

$$\lim_{M \nearrow M_Q} \frac{1}{2\pi} \int_{\partial F_M(Q)} \frac{\partial N(P, Q)}{\partial \nu} \, ds(P),$$

where $M_Q = \sup N(P, Q)$, $F_M(Q) = \{P \in R;\ N(P, Q) \geqq M\}$ and $\partial F_M(Q)$ is assumed to be complete; it is known that $\partial F_M(Q)$ is complete for almost all $M < M_Q$. The value of V at a non-minimal point Q is defined to be $\int V d\mu$ where $\mu$ is the measure which gives the canonical

1) This paper is quoted as [3] in our References.
2) This is called $\overline{\text{s}}$uperharmonic by Kuramochi and positive vollsuperharmonisch by Constantinescu and Cornea [1].

potential representation of N(P, Q). Finally, it is shown that Kuramochi's definition coincides with that of C. Constantinescu and A. Cornea [1].

For the notion of extremal length we refer to [5].

## § 1. Function $\omega_F$

We shall be interested in an open Riemann surface R. Let $K_0$ be a compact set in R bounded by a closed analytic curve and F be a closed set in $R - K_0$ with piecewise analytic boundary. In general, the relative boundary of a set in R is called piecewise analytic if it consists of analytic arcs which do not cluster in R.

It is known that there exists a unique harmonic function in $R - K_0 - F$ which takes the boundary value 0 on $\partial K_0$ and 1 on $\partial F$ and which has the smallest Dirichlet integral; cf. Theorem 11 of [4], footnote 5) of [5]. We shall denote it by $\omega_F$.

We shall state some properties of $\omega_F$ which will be needed later.

*Proposition.* (i) Let $\{F_m\}$ be a sequence of closed sets outside $K_0$ with piecewise analytic boundary, decreasing to a closed set F (which may be empty) with piecewise analytic boundary. Then $\omega_{F_m}$ decreases to a harmonic function $\omega_0$ in Dirichlet norm in $R - K_0 - F$.

(ii) Let $\alpha$ be any number such that $0 < \alpha < 1$. Then for $F_\alpha = \{P \in R - K_0;\ \omega_F(P) \geqq \alpha\} \cup F$, it holds that $\omega_F(P) = \alpha\omega_{F_\alpha}(P)$ in $R - K_0 - F$.

(iii) If $0 < \omega_F < 1$, then

$$\int_{\partial F} \frac{\partial \omega_F}{\partial \nu}\, ds = \|\omega_F\|^2$$

for almost all $\alpha$, $0 < \alpha < 1$, where the normals are drawn so that $\partial\omega_F/\partial\nu$ is non-negative everywhere on $\partial F_\alpha$.

*Proof.* (i) This is a special case of Theorem 6 of [5]. See § 5 of [5] too.

(ii) This is included in Theorem 9 of [4].[3)]

(iii) Let $\{R_n\}$ be an exhaustion of R. Let $\omega_n$ be the harmonic function in $R_n - K_o - F$, which is equal to zero on $\partial K_o$ and to 1 on the closure of $\partial F \cap R_n$ and whose normal derivative vanishes everywhere on the rest of the boundary. We see easily that $\{\omega_n\}$ converges to $\omega_F$ in Dirichlet norm and locally uniformly (cf. footnote 5) of [5]). For every $\alpha$, $0 < \alpha < 1$, we have

$$\int_{\partial F_\alpha \cap R_n} \frac{\partial \omega_n}{\partial \nu} ds = \|\omega_n\|^2$$

by Green's formula. By Fuglede's lemma (cf. [5]) there is a subsequence $\{n_k\}$ such that

$$\lim_{k\to\infty} \int_\gamma |\text{grad } (\omega_F - \omega_{n_k})| ds = 0$$

for systems $\gamma$ of locally rectifiable curves in $R' - F$ except for those belonging to a family with infinite extremal length, where $\omega_{n_k}$ is defined to be zero in $R' - F - R_{n_k}$. The extremal length of $\{\partial F_t;\ t \in E \subset (0, 1)\}$ is infinite if and only if the linear measure of E is zero. Hence for almost all t, $0 < t < 1$, it holds that

$$\lim_{k\to\infty} \left| \int_{\partial F_t} \frac{\partial \omega_F}{\partial \nu} ds - \int_{\partial F_t} \frac{\partial \omega_{n_k}}{\partial \nu} ds \right| \leqq \lim_{k\to\infty} \int_{\partial F_t} |\text{grad } (\omega_F - \omega_{n_k})| ds = 0,$$

---

3) $\partial F$ and $\partial F'$ are assumed to be analytic in [4] but Theorem 9 of [4] remains true if they are piecewise analytic.

and it follows that

$$\int_{\partial F_t} \frac{\partial \omega_F}{\partial \nu} ds = \lim_{k\to\infty} \int_{\partial F_t \cap R_{n_k}} \frac{\partial \omega_{n_k}}{\partial \nu} ds = \lim_{k\to\infty} \|\omega_{n_k}\|^2 = \|\omega_F\|^2.$$

This completes our proof.

In general, we shall call $\partial F$ *complete* if $\int_{\partial F} \partial\omega_F/\partial\nu \, ds = \|\omega_F\|^2$.

§ 2. Function N

Let $\{R_n\}$ be an exhaustion such that $K_o \subset R_1$, and set $R' = R - K_o$ and $R'_n = R_n - K_o$. Let $N_n(P, Q)$ be the function on $R'_n \times R'_n$ with the property that, as a function of P for any fixed Q, it vanishes on $\partial K_o$, it is harmonic in $R'_n$, it has the form

$$\log 1/|z(P) - z(Q)| + \text{a harmonic function}$$

in a neighborhood of Q in terms of a local parameter z, and its normal derivative vanishes everywhere on $\partial R_n$. We can see that $N_n(P, Q)$ converges locally uniformly to a function $N(P, Q)$ as a function of P outside Q and that $\| N_n - N \| \to 0$ as $n \to \infty$; see § 3 of [4].

Let $Q_k \in R'$ tend to the ideal boundary of R. If $N(P, Q_k)$ converges locally uniformly in R', $\{Q_k\}$ is called a fundamental sequence. Two fundamental sequences are said to be equivalent if the two limiting functions coincide. A *Kuramochi boundary point* Q corresponds to an equivalence class, and $N(P, Q)$ is defined to be the limiting function. We call the set of all Kuramochi boundary points the *Kuramochi boundary* of R and denote it by $\Delta_N$. By means of the distance

$$d(Q_1, Q_2) = \sup_{P\in R'_1} \left| \frac{N(P,Q_1)}{1 + N(P,Q_1)} - \frac{N(P,Q_2)}{1 + N(P,Q_2)} \right| ,$$

for $Q_1, Q_2 \in R' \cup \Delta_N$, $(R - R_1) \cup \Delta_N$ is made a compact metric space. The topology induced on $R - R_1$ coincides with the original topology.

In [4] we defined the operation $V_K$ for any positive lower semicontinuous function V in R' and for any regular compact set $K \subset R'$. If $V_k \leqq V$ for every K, V is called an SHS function.[2] Then V is superharmonic in R'. If, furthermore, V is harmonic in R', it is called an HS function. Regarding $K_o$ as corresponding to a closed disk in a parametric circle, take a sequence $\{K_m\}$ of closed sets in R which corresponds to a sequence of closed concentric disks and dicreases to $K_o$. If $V_{\partial K_m}(P)$ tends to zero as $m \to \infty$ in R', an SHS (HS resp.) function V(P) is called an $SHS_o$ ($HS_o$ resp.) function[4] in [4]. It is known (see p. 287 of [4]) that N(P, Q) is an $SHS_o$ function, where $Q \in R' \cup \Delta_N$.

For any SHS function V in R' and a closed set $F \subset R'$ with analytic boundary, we defined $V_F$ by the increasing limit of $V_{F \cap \overline{R}_n}$ in [4]. Similarly we define $V_F$ also in case F has a piecewise analytic boundary.

First we prove

*Theorem* 1. *Let* F *be a closed set in* R' *with piecewise analytic complete boundary. Then*

$$\frac{1}{2\pi} \int_{\partial F} N(Q', P) \frac{\partial \omega_F}{\partial \nu} ds(Q') = \begin{cases} \omega_F(P) & \text{if } P \in R' - F, \\ 1 & \text{if } P \in F. \end{cases}$$

*Proof.* First suppose $P \in R' - F$. Let $N_n(P, Q)$ be the function in $R'_n$ defined at the beginning of the present section. Let $\omega_n(P)$ be

4) This is called of potential type in [1].

the harmonic function in $R_n' - F$ which is equal to 0 on $\partial K_o$ and to 1 on $\partial F \cap R_n$ and whose normal derivative vanishes everywhere on $\partial R_n - \partial(F \cap R_n)$. It tends to $\omega_F(P)$ locally uniformly in R'.

By Green's formula

$$\left(\int_{\partial F \cap R_n} + \int_{\partial R_n - \partial(F \cap R_n)}\right)\left(N_n \frac{\partial \omega_n}{\partial \nu} - \omega_n \frac{\partial N_n}{\partial \nu}\right) ds = 2\pi\omega_n(P)$$

for $P \in R_n' - F$, where the normal derivatives are drawn outward with respect to $R_n' - F$. It follows that

$$2\pi\omega_n(P) = \int_{\partial F \cap R_n} N_n \frac{\partial \omega_n}{\partial \nu} ds \qquad \text{if } P \in R_n' - F.$$

It will suffice to show that the right hand side tends to $\int_{\partial F} N(\partial\omega_F/\partial\nu) ds$ as $n \to \infty$. Given $\varepsilon > 0$, we choose $p_o$ such that

$$\|\omega_F\|^2 - \int_{\partial F \cap R_{p_o}} \frac{\partial \omega_F}{\partial \nu} ds < \varepsilon \qquad \text{and} \qquad \int_{\partial F - R_{p_o}} \frac{\partial \omega_F}{\partial \nu} ds < \varepsilon.$$

We have

$$\overline{\lim_{n\to\infty}} \int_{\partial F \cap (R_n - R_{p_o})} \frac{\partial \omega_n}{\partial \nu} ds = \overline{\lim_{n\to\infty}} \left(\int_{\partial F \cap R_n} \frac{\partial \omega_n}{\partial \nu} ds - \int_{\partial F \cap R_{p_o}} \frac{\partial \omega_n}{\partial \nu} ds\right)$$

$$= \overline{\lim_{n\to\infty}} \left(\|\omega_n\|^2 - \int_{\partial F \cap R_{p_o}} \frac{\partial \omega_n}{\partial \nu} ds\right) = \|\omega_F\|^2 - \int_{\partial F \cap R_{p_o}} \frac{\partial \omega_F}{\partial \nu} ds < \varepsilon.$$

Hence

$$\overline{\lim_{n\to\infty}} \left|\int_{\partial F \cap R_n} N_n \frac{\partial \omega_n}{\partial \nu} ds - \int_{\partial F} N \frac{\partial \omega_F}{\partial \nu} ds\right|$$

$$\leqq \overline{\lim_{n\to\infty}} \left| \int_{\partial F \cap R_{p_o}} N_n \frac{\partial \omega_n}{\partial \nu} ds - \int_{\partial F \cap R_{p_o}} N \frac{\partial \omega_F}{\partial \nu} ds \right| + M \overline{\lim_{n\to\infty}} \int_{\partial F \cap (R_n - R_{p_o})} \frac{\partial \omega_n}{\partial \nu} ds$$

$$+ M \int_{\partial F - R_{p_o}} \frac{\partial \omega_F}{\partial \nu} ds \leqq 2M\varepsilon,$$

where M is a finite number such that $N_n(P, Q) \leqq M$ for any $Q \in F \cap \overline{R}_n$ and for all n. Thus

$$2\pi\omega_F(P) = \lim_{n\to\infty} \int_{\partial F \cap R_n} N_n \frac{\partial \omega_n}{\partial \nu} ds = \int_{\partial F} N \frac{\partial \omega_F}{\partial \nu} ds.$$

If P is an inner point of F, we apply Green's formula and obtain $\int_{\partial F} N(\partial\omega_F/\partial\nu)ds = 2\pi$ easily. Being a potential, $\int_{\partial F} N(\partial\omega_F/\partial\nu)ds$ is superharmonic in R'. By the mean value property it takes the value $2\pi$ on $\partial F$ too.

§ 3. Minimal points

Let A be a closed subset of $\Delta_N$. Using the metric d given at the beginning of § 2, we set

$$A(m) = \{P \in R';\ d(P, A) \leqq \tfrac{1}{m}\}.$$

At p. 289 of [4] we showed the existence of a decreasing sequence of closed neighborhoods of A in $R' \cup \Delta_N$ such that each of their intersections $\{A^{(m)}\}$ with R' has an analytic boundary in R' and $A(m) \subset A^{(m)} \subset A(m-1)$ for each m.

For an SHS function V in R' we set $V_A = \lim_{m\to\infty} V_{A^{(m)}}$. It is represented as $\int_A N(P, Q)d\mu(Q)$ on R' with some non-negative Radon measure $\mu$ supported by A; see Theorem 16 of [4]. Theorem 21 of [4] asserts that

$$\frac{1}{2\pi}\int_{\partial K_0}\frac{\partial N_{\{Q\}}(\cdot,\ Q)}{\partial\nu}\,ds = 1 \quad\text{or}\quad 0 \qquad\qquad \text{for } Q\in\Delta_N.$$

Correspondingly we shall call Q *minimal* or *non-minimal*. The set of minimal (non-minimal resp.) points will be denoted by $\Delta_1$ ($\Delta_0$ resp.). Any $SHS_0$ function is represented uniquely as $\int_{R'\cup\Delta_1} N d\mu$ (cf. Corollary of Theorem 24 of [4]); this is called a canonical representation. A point $Q\in\Delta_N$ is called *singular* if $\omega_{\{Q\}} > 0$. It belongs to $\Delta_1$ as stated in [4]. The set of all singular points will be denoted by $\Delta_S$.

Let Q be a point of $(R - K_0)\cup\Delta_N$ and set

$$F_M(Q) = \{P\in R';\ N(P,\ Q) \geqq M\}.$$

For $A = \{Q\}$ we shall write $v_m$ instead of $A^{(m)}$.

First we prove

*Lemma* 1. Let $Q\in R'\cup(\Delta_1 - \Delta_S)$. Then $\sup\limits_{P\in R'} N(P,\ Q) = \infty$.

*Proof*. Suppose $N(P,\ Q) \leqq M < \infty$ on R'. Take $\{v_m\}$ as above. It holds that $N_{v_m\cap\overline{R}_n}(P,\ Q) \leqq M\omega_{v_m\cap\overline{R}_n}(P)$ in R'. By letting $n\to\infty$ we obtain $N_{v_m}(P,\ Q) \leqq M\omega_{v_m}(P)$ in R'. We know that $N_{\{Q\}} \leqq N_{v_m} \leqq N$ and that $N_{\{Q\}} = N$ on account of the assumption $Q\in R'\cup\Delta_1$. Hence $N_{v_m} = N$ and it follows that $N(P,\ Q) \leqq M\omega_{v_m}(P) \to M\omega_{\{Q\}}(P) = 0$, because $Q\notin\Delta_S$. Namely $N(P,\ Q) \equiv 0$ at every $P\in R'$, which is impossible. Therefore $N(P,\ Q)$ is unbounded.

Let us write $M_Q$ for $\sup\limits_{P\in R'} N(P,\ Q)$. We shall prove next

*Theorem* 2. *Let* $Q\in\Delta_1$. *Then*

$$N_{F_M(Q)}(P, Q) = N(P, Q) \qquad \textit{for any } M,\ 0 < M < M_Q.$$

*Proof.* Take $\{v_m\}$ as before. If there is m such that $F_M(Q) \supset v_m$, then

$$N = N_{\{Q\}} \leqq N_{F_M(Q)} \leqq N.$$

Hence suppose that $v_m - F_M(Q) \neq \emptyset$ for all m. The function $(N(\cdot, Q))_{v_m - F_M^i(Q)}$ is an $SHS_o$ function and decreases to an $HS_o$ function as $m \to \infty$, where $F_M^i(Q) = \{P \in R';\ N(P, Q) > M\}$. This function will be denoted by $N'(P)$. It is easy to observe that $N'(P) \leqq M$. If $Q \in \Delta_1 - \Delta_S$, then it is seen as in the proof of Theorem 16 of [4] that $N'(P) = \int_{\{Q\}} N d\mu = cN(P, Q)$ in R'. Since $\sup N(P, Q) = \infty$ on R' by Lemma 1, c and hence N' must vanish. If $Q \in \Delta_S$, then $\omega_{\{Q\}}(P) = \int_{\{Q\}} N d\mu = c'N(P, Q)$. Since $\omega_{\{Q\}} > 0$, $c' > 0$. We note that $\sup \omega_{\{Q\}} = 1$, because $\omega_{\{Q\}} = (\omega_{\{Q\}})_{\{Q\}} \leqq (\sup \omega_{\{Q\}})\omega_{\{Q\}}$ by the Corollary of Theorem 19 of [4]. Hence $M_Q = \sup N = 1/c'$. We take $v_m - F_M^i(Q)$ for $F_m$ in the Proposition and denote $\lim_{m\to\infty} \omega_{F_m}$ by $\omega_o$ as there. By Theorem 9 of [4], $(\omega_{F_p})_{F_m} = \omega_{F_p}$ [3] if $m < p$. By Proposition, (i) we have $\overline{\lim}_{p\to\infty} \|(\omega_{F_p} - \omega_o)_{F_m}\| \leqq \lim_{p\to\infty} \|\omega_{F_p} - \omega_o\| = 0$. Therefore

$$0 = \lim_{p\to\infty} \|(\omega_{F_p} - \omega_o)_{F_m}\| = \lim_{p\to\infty} \|\omega_{F_p} - (\omega_o)_{F_m}\| = \|\omega_o - (\omega_o)_{F_m}\|.$$

Thus $\omega_o = (\omega_o)_{F_m}$. Consequently

$$\omega_o = (\omega_o)_{F_m} \leqq (\omega_{\{Q\}})_{F_m} \leqq Mc'\omega_{F_m} \to Mc'\omega_o \qquad \text{as } m \to \infty,$$

because $\omega_{\{Q\}} \leqq Mc'$ in $R' - F_M^1(Q)$. It follows that

$$0 \geqq (1 - Mc')\omega_o = \left(1 - \frac{M}{M_Q}\right)\omega_o \geqq 0,$$

so that $\omega_o \equiv 0$. Therefore $\lim_{m\to\infty} N_{F_m} = \lim_{m\to\infty} c'^{-1}(\omega_{\{Q\}})_{F_m} \leqq M\omega_o \equiv 0$.

In all cases we have $N' = \lim_{m\to\infty} N_{v_m - F_M^1(Q)} = 0$ and

$$N = N_{\{Q\}} \leqq \lim_{m\to\infty} N_{v_m \cap F_M(Q)} + \lim_{m\to\infty} N_{v_m - F_M^1(Q)} \leqq N_{F_M(Q)} \leqq N,$$

whence $N_{F_M(Q)} = N$.

*Theorem* 3. *Let* $Q \in \Delta_1$ *and* $0 < M < M_q$. *Then there exists* m *such that*

$$v_m \subset F_M(Q).$$

*Proof.* Suppose that there are $Q_1, Q_2, \ldots$ in $R' - F_M(Q)$ such that $Q_i \to Q$. Take M" such that $M < M'' < M_Q$. By the preceding theorem, $N(P, Q) = M''\omega_{F_{M''}(Q)}(P)$ in $R' - F_{M''}(Q)$. There is M', $M < M' < M''$, such that $\partial F_{M'}(Q)$ is complete by Proposition, (ii) and (iii). We have

$$(1) \quad \int_{\partial F_{M'}(Q)} \frac{\partial N}{\partial \nu}\, ds = M'' \int_{\partial F_{M'}(Q)} \frac{\partial \omega_{F_{M''}(Q)}}{\partial \nu}\, ds = M'' \int_{\partial K_o} \frac{\partial \omega_{F_{M''}(Q)}}{\partial \nu}\, ds$$

$$= \int_{\partial K_o} \frac{\partial N}{\partial \nu}\, ds = 2\pi.$$

By Theorem 1

$$M > N(Q_i, Q) = M'\omega_{F_{M'}(Q)}(Q_i) = \frac{M'}{2\pi} \int_{\partial F_{M'}(Q)} N(Q_i, Q') \frac{\partial \omega_{F_{M'}}}{\partial \nu}\, ds(Q')$$

$$= \frac{1}{2\pi} \int_{\partial F_{M'}(Q)} N(Q_1, Q') \frac{\partial N(Q', Q)}{\partial \nu} ds(Q').$$

Take any $M^*$, $M < M^* < M'$, and put $\varepsilon_0 = 2\pi(1 - M/M^*) > 0$. By (1) there exists $n_0$ such that

$$\int_{\partial F_{M'}(Q) \cap R_n} \frac{\partial N}{\partial \nu} ds \geqq 2\pi - \varepsilon_0 \qquad \text{for } n \geqq n_0.$$

If $N(Q_1, Q') \geqq M^*$ everywhere on $\partial F_{M'}(Q) \cap R_{n_0}$, then

$$M > \frac{1}{2\pi} \int_{\partial F_{M'}(Q) \cap R_{n_0}} N(Q', Q_1) \frac{\partial}{\partial \nu} N(Q', Q) ds \geqq \frac{M^*}{2\pi} (2\pi - \varepsilon_0) = M.$$

This is impossible. Hence there exists $P_i \in \partial F_{M'}(Q) \cap R_{n_0}$ such that $N(P_i, Q_i) < M^* < M'$. Since $\partial F_{M'}(Q) \cap \bar{R}_{n_0}$ is compact, there is a point $P_0$ of accumulation on it for $\{P_i\}$. We have

$$M' = N(P_0, Q) \leqq \overline{\lim_{i \to \infty}} N(P_i, Q_i) \leqq M^*.$$

This is absurd and our theorem is established.

*Theorem 4. Let* $Q \in \Delta_1$ *and* $0 < M < M_Q$. *Then every component of* $F_M(Q)$ *is not disjoint from* $F_{M'}(Q)$ *for any* $M'$ *such that* $M < M' < M_Q$.

*Proof.* Suppose there is a component $F$ of $F_M(Q)$ which is disjoint from $F_{M'}(Q)$. Let $f(P)$ be a function in $R' - F_{M'}$ which is equal to $N(P, Q)$ outside of $F$ and to $M$ on $F$. It belongs to $\mathcal{D}_{R'-F_{M'}(Q)}(N)$[5] and

5) See § 1 of [4] for the definition.

$$\|f\|_{R'-F_{M'}(Q)} < \|N\|_{R'-F_{M'}(Q)} = \|N_{F_{M'}(Q)}\|_{R'-F_{M'}(Q)}.$$

This contradicts the fact that $N_{F_{M'}(Q)}$ has the minimum integral among the functions of $\mathcal{D}_{R'-F_{M'}(Q)}(N)$.

*Theorem* 5. *Let* V *be an* $SHS_0$ *function, and* $Q \in \Delta_1$. *Let* $0 < M < M' < M_Q$ *and suppose that* $\partial F_M(Q)$ *and* $\partial F_{M'}(Q)$ *are complete. Then* $\int_{\partial F_M(Q)} V(\partial N/\partial\nu)ds$ *increases to* $\int_{\partial F_{M'}(Q)} V(\partial N/\partial\nu)ds$ *as* $M \nearrow M'$.

*Proof.* We begin with the case where $V(P) = N(P, P_0)$ with $P_0 \in R'$. By Theorem 1 $\int_{\partial F_M(Q)} N(Q', P_0)\partial\omega_{F_M(Q)}/\partial\nu \, ds(Q')$

$= 2\pi\omega_{F_M(Q)}(P_0)$ if $P_0 \in R' - F_M(Q)$ and $= 2\pi$ if $P_0 \in F_M(Q)$. A similar relation is true for M'. Hence

$$\frac{1}{2\pi}\int_{\partial F_{M'}(Q)-\partial F_M(Q)} N(Q', P_0)\,\frac{\partial N(Q', Q)}{\partial\nu}\,ds(Q')$$

$$= \begin{cases} 0 & \text{if } P_0 \notin F_M(Q), \\ N(P_0, Q) - M & \text{if } P_0 \in F_M(Q) - F_{M'}(Q), \\ M' - M & \text{if } P_0 \in F_{M'}(Q). \end{cases}$$

Next we consider $V(P) = \int_{R'} N d\mu$. Then

$$\frac{1}{2\pi}\int_{\partial F_{M'}(Q)-\partial F_M(Q)} V\,\frac{\partial N}{\partial\nu}\,ds = \frac{1}{2\pi}\int\left(\int_{R'} N d\mu\right)\frac{\partial N}{\partial\nu}\,ds$$

$$= \int_{F_M(Q)-F_{M'}(Q)} \{N(P, Q) - M\}d\mu(P) + \int_{F_{M'}(Q)} (M' - M)d\mu(Q) \geqq 0.$$

Finally, in the case where V(P) is harmonic in R', V(P) is expressed as a potential $\int_{\partial R_n} N d\mu_n$ on $R_n'$ for each n and $\mu_n(R)$ is bounded (cf. [4], Theorem 13). We have

$$\frac{1}{2\pi}\int_{\partial F_{M'}(Q)-\partial F_M(Q)} V\,\frac{\partial N}{\partial \nu}\,ds$$

$$= \lim_{n\to\infty}\left[\int_{F_M(Q)-F_{M'}(Q)}\{N(P,\ Q)-M\}d\mu_n(P) + (M'-M)\mu_n(F_{M'}(Q))\right] \geqq 0$$

and

$$\frac{1}{2\pi}\int_{\partial F_{M'}(Q)-\partial F_M(Q)} V\,\frac{\partial N}{\partial \nu}\,ds \leqq (M' - M)\lim_{n\to\infty}\mu_n(R') < \infty.$$

Consequently, for any $V = \int_{R'\cup\Delta_N} N d\mu$, $\int_{\partial F_M(Q)} V(\partial N/\partial\nu)ds$ increases to $\int_{\partial F_{M'}(Q)} V(\partial N/\partial\nu)ds$ as $M \nearrow M'$.

We shall write $\int V dN^*$ for $\int V(\partial N/\partial\nu)ds$ in what follows.

§ 4. Values of an $SHS_o$ function on $\Delta_N$

Let V(P) be an $SHS_o$ function and Q be a point of $\Delta_1$. We define V(Q) by

$$\lim_{M\nearrow M_Q}\frac{1}{2\pi}\int_{\partial F_M(Q)} V(P)dN^*(P,\ Q)$$

as M is chosen so that $\partial F_M(Q)$ is complete. The limit is an increasing limit by Theorem 5. Before determining the value of V at $Q \in \Delta_o$, we shall prove that our definition of $V$ on $\Delta_1$ coincides with that of Constantinescu-Cornea [1]. They considered first the following potential representation of $V_K$:

$$V_K(P) = \int_K N(P, Q)d\mu_K(Q) \qquad \text{in } R'$$

for a regular compact set $K \subset R'$. Since $N(P, Q)$ is continuous on the product space $K \times (R' \cup \Delta_N - K)$, $V_K(P)$ is continuous on $R' \cup \Delta_N - K$. Also it is lower semicontinuous on $R' \cup \Delta_N$. They defined the value of $V$ on $\Delta_N$ by

$$\tilde{V}(P) = \sup_K V_K(P) = \lim_{K \nearrow R'} V_K(P).$$

Naturally $\tilde{V}(P)$ is lower semicontinuous on $R' \cup \Delta_N$.

Let us prove

*Lemma* 2. At any $Q \in \Delta_1$ and for any $M$, $0 < M < M_Q$, such that $\partial F_M(Q)$ is complete, it holds that $\int_{\partial F_M(Q)} V_K dN^* \leqq 2\pi V_K(Q)$. The equality holds if $M$ is sufficiently close to $M_Q$.

*Proof*. The inequality is true because $V_K$ is an $SHS_o$ function. Next choose $M < M_Q$ so that $\partial F_M(Q)$ is complete and $F_M(Q) \cap K = \emptyset$. We have

$$\frac{1}{2\pi}\int_{\partial F_M(Q)} V_K dN^* = \int_K \left(\frac{1}{2\pi}\int_{\partial F_M(Q)} N(P, Q')dN^*(P, Q)\right) d\mu_K(Q')$$

$$= \int_K N(Q, Q')d\mu_K(Q') = V_K(Q).$$

This completes our proof.

Now we prove $V(Q) = \tilde{V}(Q)$ for any $Q \in \Delta_1$. Since $V_K(P) \leqq V(P)$ in $R'$,

$$V(Q) \geq \lim_{M \nearrow M_Q} \frac{1}{2\pi} \int_{\partial F_M(Q)} V_K(P) dN^*(P, Q) = V_K(Q)$$

by Lemma 2. It follows that $V(Q) \geq \tilde{V}(Q)$. On the other hand,

$$V(Q) = \lim_{M \nearrow M_Q} \frac{1}{2\pi} \int_{\partial F_M(Q)} \lim_{K \nearrow R'} V_K(P) dN^*(P, Q)$$

$$= \lim_{M \nearrow M_Q} \lim_{K \nearrow R'} \frac{1}{2\pi} \int_{\partial F_M(Q)} V_K(P) dN^*(P, Q)$$

$$\leq \lim_{M \nearrow M_Q} \lim_{K \nearrow R'} V_K(Q) = \tilde{V}(Q)$$

again by Lemma 2. Thus $V(Q) = \tilde{V}(Q)$ on $\Delta_1$.

It is known that $\Delta_o$ is an $F_\sigma$-set (see Theorem 22 of [4]). As a function on $\Delta_1$, $V(Q)$ is lower semicontinuous. Given $Q \in \Delta_o$, we consider the canonical representation

$$N(P, Q) = \int_{\Delta_1} N(P, Q') d\mu(Q') \qquad \text{for } P \in R',$$

and define $V(Q)$ by $\int_{\Delta_1} V(Q') d\mu(Q')$. We shall prove

*Theorem 6.* $V(Q) = \tilde{V}(Q)$ *everywhere on* $\Delta_N$.

*Proof.* It suffices to prove the equality for $Q \in \Delta_o$. We have

$$V(Q) = \int_{\Delta_1} V(Q') d\mu(Q') = \int_{\Delta_1} \tilde{V}(Q') d\mu(Q') = \lim_{K \nearrow R'} \int_{\Delta_1} V_K(Q') d\mu(Q')$$

$$= \lim_{K \nearrow R'} \int_{\Delta_1} \int_K N(P, Q') d\mu_K(P) d\mu(Q')$$

$$= \lim_{K \nearrow R'} \int_K \left\{ \int_{\Delta_1} N(P, Q') d\mu(Q') \right\} d\mu_K(P)$$

$$= \lim_{K \nearrow R'} \int_K N(P, Q) d\mu_K(P) = \tilde{V}(Q).$$

Finally we prove

*Theorem* 7. (i) $N(P, Q) = N(Q, P)$ *for any* $P, Q \in R' \cup \Delta_N$.

(ii) $N(Q, Q) = M_Q$ *for* $Q \in \Delta_1$.

(iii) *The canonical representation of* $V(P)$ *valid on* $R'$ *is also extended to* $\Delta_N$. *Namely,*

$$V(P) = \int_{R' \cup \Delta_1} N(P, Q) d\mu(Q) \qquad \text{on } R' \cup \Delta_N.$$

*Proof.* (i) We begin with the case where $P \in R'$ and $Q \in \Delta_1$. Let M be a large mumber such that $P \notin F_M(Q)$ and $\partial F_M(Q)$ is complete. Then by Theorem 1

$$N(P, Q) = \frac{1}{2\pi} \int_{\partial F_M(Q)} N(Q', P) dN^*(Q', Q) \nearrow N(Q, P).$$

If $P \in R'$ and $Q \in \Delta_o$,

$$N(P, Q) = \int_{\Delta_1} N(P, Q') d\mu(Q') = \int_{\Delta_1} N(Q', P) d\mu(Q') = N(Q, P).$$

If $P \in \Delta_1$ and $Q \in \Delta_N$,

$$N(P, Q) = \lim_{M \nearrow M_P} \frac{1}{2\pi} \int_{\partial F_M(P)} N(Q', Q) dN^*(Q', P)$$

$$= \lim_{M \nearrow M_P} \frac{1}{2\pi} \int_{\partial F_M(P)} N(Q, Q') dN^*(Q', P) = N(Q, P).$$

In case $P \in \Delta_o$ and $Q \in \Delta_N$, we represent $N(P', P)$ by $\int_{\Delta_1} N(P', Q') d\mu(Q')$ and have

$$N(P, Q) = \int_{\Delta_1} N(Q', Q) d\mu(Q') = \int_{\Delta_1} N(Q, Q') d\mu(Q') = N(Q, P).$$

(ii) $N(Q, Q) = \lim_{M \nearrow M_Q} \frac{1}{2\pi} \int_{\partial F_M(Q)} N(Q', Q) dN^*(Q', Q) = \lim_{M \nearrow M_Q} M = M_Q.$

(iii) Represent V(P) by $\int_{R'\cup\Delta_1} N(P, Q')d\mu(Q')$ in R'.

Let $Q \in \Delta_1$. Then by definition

$$V(Q) = \lim_{M \nearrow M_Q} \int_{\partial F_M(Q)} V(P)dN^*(P, Q)$$

$$= \lim_{M \nearrow M_Q} \int_{\partial F_M(Q)} \int_{R'\cup\Delta_1} N(P, Q')d\mu(Q')dN^*(P, Q)$$

$$= \lim_{M \nearrow M_Q} \int_{R'\cup\Delta_1} \int_{\partial F_M(Q)} N(P, Q')dN^*(P, Q)d\mu(Q')$$

$$= \int_{R'\cup\Delta_1} \left(\lim_{M \nearrow M_Q} \int_{\partial F_M(Q)} N(P, Q')dN^*(P, Q)\right)d\mu(Q')$$

$$= \int_{R'\cup\Delta_1} N(Q, Q')d\mu(Q').$$

Next let $Q \in \Delta_0$, and $N(P, Q) = \int_{\Delta_1} N(P, Q')d\nu(Q')$. Then

$$V(Q) = \int_{\Delta_1} V(Q')d\nu(Q') = \int_{\Delta_1} \int_{R'\cup\Delta_1} N(Q', P')d\mu(P')d\nu(Q')$$

$$= \int_{R'\cup\Delta_1} N(P', Q)d\mu(P') = \int_{R'\cup\Delta_1} N(Q, Q')d\mu(Q').$$

Finally we remark that the values of any SHS function V are defined on $\Delta_N$. To show it, consider a ring domain D in R' partly bounded by $\partial K_0$ and replace V in D by the Dirichlet solution for the boundary function 0 on $\partial K_0$ and V on $\partial D \cap R'$. The resulting function on R' is an $SHS_0$ function and its values on $\Delta_N$ are well-defined. Naturally we regard these values as those of V.

*References*

[1] C. Constantinescu and A. Cornea: Ideale Ränder Riemannscher Flächen, Berlin-Göttingen-Heidelberg, 1963.

[2] Z. Kuramochi: Mass distributions on the ideal boundaries of abstract Riemann surfaces, II, Osaka Math. J., 8 (1956), 145-186.

[3] Z. Kuramochi: Potentials on Riemann surfaces, J. Fac. Sci. Hokkaido Univ. Ser. I, 16 (1962), 5-79.

[4] M. Ohtsuka: An elementary introduction of Kuramochi boundary, J. Sci. Hiroshima Univ. Ser. A-I Math., 28 (1964), 271-299.

[5] M. Ohtsuka: Dirichlet principle on Riemann surfaces, J. Analyse Math., 19 (1967), 295-311.

Department of Mathematics,
Faculty of Science,
Hiroshima University

# VI. A CONDITION FOR EACH POINT OF THE KURAMOCHI BOUNDARY TO BE OF HARMONIC MEASURE ZERO

Kikuji MATSUMOTO

1. Let R be an open Riemann surface belonging to $O_{HD} - O_G$. Then its Kuramochi boundary contains precisely one point with positive harmonic measure. It is well-known that such a Riemann surface cannot be represented as any bounded-sheeted covering surface over the complex plane and so it is not of finite genus. The same is true for every open Riemann surface possessing at least one Kuramochi boundary point with positive harmonic measure. In fact, recall the following theorem due to Kuramochi, Constantinescu and Cornea (see [1]):

*Theorem. Let* R *be an open Riemann surface whose Kuramochi boundary contains at least one point with positive harmonic measure. Then*

(1) *it belongs to the class* $O_{AD}$ *and*

(2) *for any compact subset* K *of* R *such that* R - K *is connected, the Kuramochi boundary of* R - K *has also at least one point with positive harmonic measure.*

Suppose that R can be represented as a bounded-sheeted covering surface. Let N be the maximum covering number and $w_0$ a point covered by R precisely N times. Here we may assume that $w_0$ is the point at infinity and there is no branch point over it. Then for sufficiently large $r > 0$, the closed disc $|w| \geqq r$ is covered by just N discs and hence the part K of R lying over this disc is compact and has the connected complement. By the above theorem there is no nonconstant analytic function with finite Dirichlet integral on R - K, while the projection restricted to R - K has a finite Dirichlet integral. Contradiction.

Therefore if we can find properties or quantities representing nearness of an open Riemann surface to Riemann surfaces being representable as a bounded-sheeted covering surface or being of finite genus, then it will be possible to give some condition for each point of its Kuramochi boundary to be of harmonic measure zero. Recently Nakai has given such a condition "*of almost finite genus*" and proved

*Theorem* (Nakai [5]). *The Kuramochi boundary of any open Riemann surface of almost finite genus has no point with positive harmonic measure.*

Our aim is to give another condition representing the nearness by using the operations I and E, which were introduced by Kuramochi [3] and Heins [2], and to make Nakai's theorem clear from our view point.

The main results given here were proved in [4]. But, by giving two new lemmas (the 2nd and the 3rd lemmas in § 6), the proof becomes very clearer and furthermore we can give in § 7 a concrete criterion for each Martin boundary point to be of harmonic measure zero.

2. Let R be a Riemann surface and $G_i$ with $i = 1, 2, \ldots, n \leqq +\infty$ be domains on R with smooth relative boundary $\partial G_i$ clustering nowhere in R and being disjoint by pairs. For the union G of these domains $G_i$ and a positive harmonic function u on R we denote by $I_G(u)$ the upper envelope of all the nonnegative subharmonic functions on G dominated by u and vanishing continuously on $\partial G$. A nonnegative harmonic function U on G is called admissible if it vanishes continuously on $\partial G$ and if there is at least one positive superharmonic function on R dominating U on G. For an admissible U we denote by $E_G(U)$ the lower envelope of all the positive superharmonic functions on R dominating U on G. Then it is known that these operations have the following properties (see [2] and [3]).

(1) $I_G$ and $E_G$ are additive.

(2) $I_G E_G$ is the identity, that is, for any admissible U

$$I_G(E_G(U)) = U.$$

(3) Let v be a positive harmonic function on R. If there exists an admissible U on G such that $E_G(U) \geqq v$ on R, then

$$v = E_G(I_G(v)).$$

(4) Let $U_k$ $(k = 1, 2, \ldots)$ be a monotone sequence of admissible functions on G with limit U being admissible. Then

$$E_G(U) = \lim_{k\to\infty} E_G(U_k).$$

The result to be established is the following theorem.

*Theorem* ([4]). *If*

(i) *each* $G_i$ *is of finite genus or more generally representable as a bounded-sheeted covering surface over the complex plane and*

(ii) $I_{G_i}(1) \in HD(G_i)$ *with* $i = 1, \ldots, n$ *and* $E_G(I_G(1)) = 1$,

*then each point of the Kuramochi boundary of* R *is of harmonic measure zero.*

3. To prove the theorem we need a simple lemma. HP is the class of harmonic functions each of which is representable as difference of two nonnegative harmonic functions. HP $\supset$ HD and we denote by MHD the smallest monotone class containing the class HD, where a subclass of HP is called monotone if it contains all the limit-functions of its monotone sequences. We say that a positive MHD-function u is MHD-minimal if each positive MHD-minorant of u is proportional to u. Now we prove

*Lemma.* Let u be an MHD-minimal function on R. If there exists an admissible U on G with its restriction $U_i$ to $G_i$ belonging to the class $HD(G_i)$ for every $i$, $1 \leqq i \leqq n$, such that

$$E_G(U) \geqq u,$$

then there is precisely one $G_i$, say $G_1$, such that $I_{G_1}(u) > 0$ and is MHD-minimal on $G_1$.

*Proof.* The existence of an admissible $U$ on $G$ with $E_G(U) \geqq u$ implies that $E_G(I_G(u)) = u$, and hence there is at least one $G_i$, say $G_1$, with $I_{G_1}(u) > 0$, since $I_G(u) = I_{G_i}(u)$ on each $G_i$. $U \geqq I_{G_1}(u)$ and $u \geqq u \underset{G_1}{\wedge} U \geqq I_{G_1}(u)$ on $G_1$, where for two harmonic functions $v$ and $v'$ on $G_1$ we denote by $v \underset{G_1}{\wedge} v'$ the greatest harmonic minorant of $\min(v, v')$ on $G_1$ if it exists. Obviously $u \underset{G_1}{\wedge} U$ vanishes continuously on $\partial G_1$, so that $u \underset{G_1}{\wedge} U = I_{G_1}(u)$. Hence $I_{G_1}(u)$ is an MHD-function on $G_1$.

Now let $W$ be a positive MHD-function on $G_1$ dominated by $I_{G_1}(u)$. Then $E_{G_1}(W)$ is a positive MHD-function on $R$ and dominated by $u$, whence we have $E_{G_1}(W) = cu$ with $c$, $0 < c \leqq 1$. We have $W = I_{G_1}(E_{G_1}(W)) = cI_{G_1}(u)$ and see that $I_{G_1}(u)$ is MHD-minimal on $G_1$.

Let $V$ be the harmonic function on $G$ defined by

$$V = \begin{cases} I_{G_1}(u) & \text{on } G_1 \\ 0 & \text{on } G_i \ (i = 2, \ldots, n). \end{cases}$$

On considering $I_{G_1}(u)$ as $W$ in the above we see that $E_G(V) = E_{G_1}(I_{G_1}(u)) = cu$ with $c$, $0 < c \leqq 1$. Hence $I_G(u) = c^{-1} I_G(E_G(V)) = c^{-1} V$ so that

$I_{G_i}(u) = 0$ on $G_i$ with $i = 2, \ldots, n$. The proof is now complete.

4. *Proof of the theorem.* First we recall the following result due to Constantinescu and Cornea (see [1]).

In order that the Kuramochi boundary of an open Riemann surface R has at least one point with positive harmonic measure, it is necessary and sufficient that there is at least one MHD-minimal function on R, $R \in U_{HD}$ in notation.

We shall prove the theorem by contradiction. Suppose that our theorem is false. Then our Riemann surface admits an MHD-minimal function u. It is known that u is bounded, so that we assume $0 < u < 1$. Set $U = I_G(1)$. Then the restriction $U_i = I_{G_i}(1)$ of U to $G_i$ belongs to the class $HD(G_i)$ for every i, $1 \leqq i \leqq n$ and $E_G(U) = E_G(I_G(1)) = 1 \geqq u$. Therefore this U satisfies all conditions of the above lemma and so we can conclude that there is precisely one $G_i$ with $I_{G_i}(u) > 0$ and being MHD-minimal on $G_i$. On the other hand, by our hypothesis, $G_i$ is of finite genus or representable as a bounded-sheeted covering surface and hence admits no MHD-minimal function. This contradiction proves the theorem.

5. We shall now make Nakai's theorem clear from our view point. First we shall explain Nakai's concept "of almost finite genus."

Let R be a Riemann surface. We denote by $[C_1, C_2]$ a pair of mutually disjoint simple closed curves $C_1$ and $C_2$ on R satisfying the following two conditions:

(1) $C_1$ and $C_2$ are dividing cycles of R, i.e. the open set $R - C_i$ ($i = 1, 2$) consists of two components,

(2) the union of $C_1$ and $C_2$ is the boundary of a relatively compact domain $(C_1, C_2)$ of R such that $(C_1, C_2)$ is of genus one.

We say that two such pairs $[C_1, C_2]$ and $[C'_1, C'_2]$ are equivalent if there exists a third pair $[C''_1, C''_2]$ such that $(C_1, C_2) \quad (C'_1, C'_2) \supset (C''_1, C''_2)$, or if there exists a chain of pairs $[C_1, C_2]$, $[C_1^{(1)}, C_2^{(2)}], \ldots, [C_1^{(n)}, C_2^{(n)}]$, $[C'_1, C'_2]$ such that each pair of this chain is equivalent to its next one in the above sense. Then this relation is actually an equivalence relation, so that we divide the totality of these pairs $[C_1, C_2]$ into equivalence classes. Calling each equivalence class H a handle of R, we observe that R has at most a countable number of handles.

An annulus A in R is said to be associated with a handle H of R, $A \in H$ in notation, if there exists a representative $[C_1, C_2]$ of H such that $\bar{A} \subset (C_1, C_2)$ and each boundary component of the relative boundary of A rounds the hole of $(C_1, C_2)$, that is,

(3) each boundary component of A is not a dividing cycle of the domain $(C_1, C_2)$.

We say that a Riemann surface R is of almost finite genus, if there exists a sequence $\{A_n\}$ of annuli in R satisfying

(4) $A_n \in H_n$, where $\{H_n\}$ is the totality of handles in R,

(5) $A_n \cap A_m = \emptyset$ if $n \neq m$,

(6) $\sum_n 1/\text{mod } A_n < +\infty$,

where mod $A_n$ is the harmonic modulus of the annulus $A_n$. Of course any Riemann surface of finite genus is of almost finite genus.

6. We shall show that Nakai's theorem can be obtained as a corollary of our theorem. We denote by $\gamma_n$ the closed Jordan curve in $A_n$ dividing it into two annuli $A_{n,1}$ and $A_{n,2}$ such that mod $A_{n,1}$ = mod $A_{n,2} = 2^{-1}$ mod $A_n$. Since $G = R - \bigcup_n \gamma_n$ is a subregion on R *of planar character*, it is enough for us to show that $I_G(1) \in HD(G)$

and $E_G(I_G(1)) = 1$ if $R \notin O_G$ and $\sum_n 1/\text{mod } A_n < +\infty$.

*Lemma.* If $R \notin O_G$ and $\sum_n 1/\text{mod } A_n < +\infty$, then $I_G(1) \in HD(G)$.

*Proof.* Let $w_n(p)$ be the continuous function on R such that

$$w_n(p) = \begin{cases} \text{harmonic} & \text{on } A_n - \gamma_n \\ 1 & \text{on } \gamma_n \\ 0 & \text{on } R - A_n, \end{cases}$$

and let w(p) denote the least harmonic majorant of $\sum_n w_n(p)$ on G. Then

$$w(p) + I_G(1)(p) \equiv 1 \text{ on } G$$

and

$$D(w) \leqq \sum_n D(w_n) = 8\pi \sum_n 1/\text{mod } A_n < +\infty,$$

where D(u) is the Dirichlet integral of u taken over R. Hence $I_G(1) \in HD(G)$.

*Lemma.* If $R \notin O_G$ and the harmonic measures $\omega_n(p)$ of $\gamma_n$ with respect to R satisfy the condition $\sum_n \omega_n(p) < +\infty$ on R, then $E_G(I_G(1)) = 1$.

*Proof.* Since $\omega_n(p)$ is a Green potential on R for each n and $\sum_n \omega_n(p) < +\infty$, $\sum_n \omega_n(p)$ is also a Green potential on R. On the other hand the nonnegative function $v = 1 - E_G(I_G(1))$ on R satisfies that

$$v \leqq 1 - I_G(1) \leqq \sum_n \omega_n.$$

Hence $v \equiv 0$, i.e. $E_G(I_G(1)) = 1$, because any Green potential admits no positive harmonic minorant on R.

*Lemma.* If $R \notin O_G$ and $\sum_n 1/\text{mod } A_n < +\infty$, then $\sum_n \omega_n(p) < +\infty$

on R.

*Proof.* Let $\mu_n$ be the measure associated with the potential $\omega_n(p)$. Then it is supported by $\gamma_n$ and $\mu_n(\gamma_n) = \int_{\partial A_n} \frac{\partial \omega_n}{\partial \nu} ds$ $\leq \int_{\partial A_n} \frac{\partial w_n}{\partial \nu} ds = D(w_n) = 8\pi/\text{mod } A_n$. Therefore the measure $\mu$ defined by $\mu = \mu_n$ on $\gamma_n$ for each n has a finite total mass, that is, $\mu(R) = \sum_n \mu_n(\gamma_n) \leq 8\pi \sum_n 1/\text{mod } A_n < +\infty$, and so the potential $\omega(p)$ by this $\mu$ is finite on R. Hence it follows that

$$\sum_n \omega_n(p) = \sum_n \int G(p, q) d\mu_n(q) = \int G(p, q) d\mu(q) = \omega(p) < +\infty.$$

7. In the Martin case, we can prove the following theorem.

*Theorem* ([4]). *If*

(i) *each* $G_i$ *is of finite genus or more generally representable as a bounded-sheeted covering surface over the complex plane and*

(ii) $E_G(I_G(1)) = 1$,

*then each point of the Martin boundary of* R *is of harmonic measure zero.*

Recalling the second lemma in the preceding section, we have immediately the following

*Corollary. If* $\sum_n \omega_n(p) < +\infty$ *on* R, *then the Martin boundary of* R *has no point with positive harmonic measure.*

*References*

[1] C. Constantinescu and A. Cornea: Ideale Ränder Riemannscher Flächen, Berlin-Göttingen-Heidelberg, 1963.

[2] M. Heins: On the Lindelöfian principle, Ann. of Math., 61 (1955), 440-473.

[3] Z. Kuramochi: Relations between harmonic dimensions, Proc. Japan Acad., 30 (1954), 576-580.

[4] K. Matsumoto: Analytic functions on some Riemann surfaces, II. Nagoya Math. J., 23 (1963), 153-164.

[5] M. Nakai: Genus and classification of Riemann surface, Osaka Math. J., 14 (1962), 153-180.

Mathematical Institute,
Nagoya University

# VII. EXTREMAL LENGTH AND KURAMOCHI BOUNDARY OF A SUBREGION OF A RIEMANN SURFACE

Tatsuo FUJI'I'E

## Introduction

On a Riemann surface R we consider a family $\Gamma$ of locally rectifiable curves,[1)] c, and a class $\Phi$ of non negative covariants, $\rho$, which satisfy $\iint_R \rho^2 dxdy \leqq 1$ and for which $\int_c \rho ds$ are determined ($\leqq \infty$) for every curve c of $\Gamma$. We define the extremal length $\lambda_\Gamma$ of the family $\Gamma$ as $\left(\sup_{\rho\in\Phi} \inf_{c\in\Gamma} \int_c \rho ds\right)^2$, and call each $\rho$ of $\Phi$ admissible for the problem of extremal length $\lambda_\Gamma$.

Z. Kuramochi constructed a function N(z, P), named N-Green's function, on R - K (K is a compact disk on R) and, using this function, compactified R by the method of R. S. Martin. Since this function N(z, P) has a finite Dirichlet integral D(N(z, P)) over R - K outside a neighborhood $\Delta$ of the pole P of N(z, P), $|\text{grad } N(z, P)|/\sqrt{D(N(z, P))}$ is admissible for the problem of extremal length of a family of locally rectifiable curves in R - K - $\Delta$. This fact leads us to study various problems concerning the Kuramochi boundary. For instance, the following proposition is known ([3], [4], [1]).

*Proposition* 1. *Every curve, which starts from a point of* R *and tends to the boundary, converges to a point of the Kuramochi boundary except for curves belonging to a family whose extremal length is infinite.*

In this paper we consider a subregion D of a Riemann surface R, whose relative boundary $\partial D$ consists of at most a countable number of

1) In this paper all curves are assumed to be locally rectifiable. A curve in an open subset of R is said to tend to the boundary if, for any compact subset K, an end part of the curve is disjoint from K.

analytic curves (compact or non compact) which do not cluster in R. Let $\{R_n\}$ be a regular exhaustion of R and $D_n$ be the component of $R_n \cap D$ which contains a fixed point P of D. Then, $\{D_n\}$ form an exhaustion of D, and we say $\{D - D_n\}$ determines the ideal boundary of D.

With Kuramochi [2], we consider the function $N'_n(z, P)$ in $D_n$, which is determined by the following properties:

1. $N'_n(z, P)$ is harmonic in $D_n$ except at P. In a neighborhood $\Delta(P)$ of P, $N'_n(z, P) = -\log|z - P| + u_n(z)$ with a harmonic function $u_n(z)$.

2. $N'_n(z, P) = 0$ on the closure of $\partial D \cap R_n$.

3. $\frac{\partial N'_n(z, P)}{\partial n} = 0$ on the rest of $\partial D_n$.

Then $\{N'_n(z, P)\}$ converges to a harmonic function $N'(z, P)$ uniformly on every compact set in $D - \{P\}$. It is expressed as $-\log|z - P| + u(z)$ in a neighborhood $\Delta(P)$ of P with harmonic u(z), and vanishes on $\partial D$. Furthermore $D_{D_n}(N'_n - N') \underset{n\to\infty}{\longrightarrow} 0$ and $\lim_{n\to\infty} D_{D_n - \Delta(P)}(N'_n) < \infty$, so that the Dirichlet integral $D_{D-\Delta(P)}(N'(z, P))$ is finite. We denote by G(z, P) Green's function of D with pole at P.

*Definition* ([2]). D is said to have an *ideal boundary of positive capacity* when $N'(z, P) > G(z, P)$.
This property is independent of the choice of the pole P.

When D has an ideal boundary of positive capacity, we compactify D by making use of the function $N'(z, P)$ by the same method as in the case of a whole Riemann surface. We shall denote by $B_D$ the boundary thus obtained. To each point $P \in B_D$ a function $N'(z, P)$ corresponds. Every point P of $B_D$ with positive $N'(z, P)$ will be called a Kuramochi ideal boundary point of D.

Like Proposition 1 we can prove

*Proposition* 1'. *Every curve, which starts from a point of* D *and tends to the boundary of* D, *converges to a point of* $B_D$ *except for those belonging to a family of infinite extremal length.*

We are going to characterize the property to have "positive capacity" by extremal length.

## § 1. A criterion for a subregion to have an ideal boundary of positive capacity

Let $K_r$ be a parametric disk with radius r, centered at P and contained in D. Let $U_{r,n}$ be the harmonic function in $D_n - K_r$ satisfying the following conditions.

$$\begin{cases} U_{r,n} = -\log r & \text{on } \partial K_r, \\ U_{r,n} = 0 & \text{on the closure of } \partial D \cap R_n, \\ \dfrac{\partial U_{r,n}}{\partial n} = 0 & \text{on the rest of } \partial D_n. \end{cases}$$

As $n \to \infty$, $U_{r,n}$ converges uniformly on every compact set in $D - K_r + \partial K_r$ to a limit function $U_r$ and the Dirichlet integral $D_r(U_r)$ over $D - K_r$ is equal to $\lim_{n\to\infty} D_{D_n - K_r}(U_{r,n}) < \infty$.

Let $\lambda_r$ be the extremal length of the family of curves which join $\partial K_r$ and $\partial D$ in D, and $\lambda_{r,n}$ be the extremal length of the family of curves which join $\partial K_r$ and $\partial D \cap R_n$ in $D_n$. Then we have

*Proposition* 2. $\lambda_{r,n}$ *is monotone decreasing when* n *increases and*

$$(1) \qquad \lambda_r = \lim_{n\to\infty} \lambda_{r,n} = \lim_{n\to\infty} \frac{(\log r)^2}{D_{D_n - K_r}(U_{r,n})} = \frac{(\log r)^2}{D_r(U_r)}.$$

*Proof.* Evidently $\lambda_r \leqq \lambda_{r,n}$. Take $|\text{grad } U_r|(D_r(U_r))^{-1/2}$ as $\rho$. Since $U_r$ vanishes on $\partial D$, $\int_c \rho ds \geqq (-\log r)(D_r(U_r))^{-1/2}$ holds for every c joining $\partial K_r$ and $\partial D$ in D. Therefore

$$\lambda_r \geqq \frac{(\log r)^2}{D_r(U_r)} .$$

The equality

$$\lambda_{r,n} = \frac{(\log r)^2}{D_{D_n - K_r}(U_{r,n})}$$

is well-known and (1) is derived on account of the relation

$$\lim_{n\to\infty} D_{D_n - K_r}(U_{r,n}) = D_r(U_r).$$

Remark 1. Let $\lambda_r'$ be the extremal length of the family of curves which start from $\partial K_r$ and along which $U_r$ tends to zero. We see as above $\lambda_r' \geqq (\log r)^2/D_r(U_r)$ and the equality $\lambda_r = \lambda_r'$ follows.

Remark 2. Let r be a small number such that $\{z;\ N'(z, P) \geqq -\log r\}$ is compact in D. We may take this set as $K_r$. Then, according to Remark 1, $\lambda_r$ is equal to the extremal length of the family of curves which start from $\partial K_r$ and along which $N'(z, P)$ tends to zero.

On the other hand,

$$\begin{aligned} D_r(U_r) &= D_r(N' + U_r - N') \\ &= D_r(N') - 2D_r(N', N' - U_r) + D_r(N' - U_r). \end{aligned}$$

By Green's formula and by the fact that $N_n'(z, P) - N'(z, P)$ converges to 0 uniformly on every compact set in D, we have

$$D_r(N') = \lim_{n\to\infty} D_{D_n - K_r}(N_n') = -2\pi\log r + 2\pi u(0) + \varepsilon_1(r),$$

$$D_r(N' - U_r) = \varepsilon_2(r)$$

and

$$D_r(N', N' - U_r) = \int_{\partial K_r} u d\theta - r\int_{\partial K_r} u \frac{\partial u}{\partial r} d\theta = 2\pi u(0) + \varepsilon_3(r).$$

Therefore,

$$D_r(U_r) = -2\pi \log r - 2\pi u(0) + \varepsilon(r),$$

where $\varepsilon(r) \to 0$ when $r \to 0$.

Here, according to K. Strebel, we define the extremal radius $R(\partial D)$ of $\partial D$ measured at P by

$$\lim_{r \to o} re^{2\pi\lambda_r}.$$

Then, from the above result

$$R(\partial D) = e^{u(0)}.$$

Next, we consider the family $\Sigma_r$ of curves which start from $\partial K_r$ and tend to the boundary of D. We define R(bdy D) by $\lim_{r \to o} re^{2\pi\mu_r}$, where $\mu_r$ is the extremal length of $\Sigma_r$.

Let G(z, P) be Green's function of D with pole at P, and let $G(z, P) = -\log r + h(z)$ in a neighborhood of P. Then, we can show

$$R(\text{bdy } D) = e^{h(0)}$$

by the same method as above.

Comparing two extremal radii, we have the following theorem.

*Theorem. A subregion* D *has an ideal boundary of positive capacity if and only if* $R(\partial D) > R(\text{bdy } D)$.

§ 2. Curves converging to points of Kuramochi boundary of a subregion

Let $\Gamma_1$ ($\Gamma_2$ resp.) consist of curves in D which start from $\partial K_r$ and tend to the boundary of D (terminate at points of $\partial D$ resp.). The theorem shows that if D has an ideal boundary of positive capacity and if r is small, then the extremal length of the family $\Gamma = \Gamma_1 - \Gamma_2$ is finite. Let $\Gamma_3$ be the subfamily of $\Gamma_1$ consisting of curves along which N'(z, P) has limit 0. By Remark 2, $\lambda_{\Gamma_2} = \lambda_{\Gamma_3}$. Hence, putting $\Gamma_o = \Gamma_1 - \Gamma_3$, $\lambda_{\Gamma_o}$ is finite. On account of Proposition 1', each curve of $\Gamma_o$ converges to a Kuramochi ideal boundary point of D except for curves belonging to a family of infinite extremal length. From this fact and Proposition 1, we conclude that each curve of $\Gamma_o$ converges not only to a Kuramochi ideal boundary point of D but also to a Kuramochi boundary point of R except for curves belonging to a family of infinite extremal length.

### *References*


[1] T. Fuji'i'e, Extremal length and Kuramochi boundary, J. Math. Kyoto Univ., 4 (1964), 149-159.

[2] Z. Kuramochi, Singular points of Riemann surfaces, J. Fac. Sci. Hokkaido Univ. Ser. I, 16 (1962), 80-148.

[3] F-Y. Maeda, Notes on Green lines and Kuramochi boundary of a Green space, J. Sci. Hiroshima Univ. Ser. A-I Math., 28 (1964), 59-66.

[4] M. Ohtsuka, On limits of BLD functions along curves, ibid., 67-70.


Ritsumeikan University

Offsetdruck: Julius Beltz, Weinheim/Bergstr.

# Lecture Notes in Mathematics

**Bisher erschienen/Already published**

Vol. 1: J. Wermer, Seminar über Funktionen-Algebren. IV, 30 Seiten. 1964. DM 3,80 / $ 0.95

Vol. 2: A. Borel, Cohomologie des espaces localement compacts d'après J. Leray. IV, 93 pages. 1964. DM 9,– / $ 2.25

Vol. 3: J. F. Adams, Stable Homotopy Theory. 2nd. revised edition. IV, 78 pages. 1966. DM 7,80 / $ 1.95

Vol. 4: M. Arkowitz and C. R. Curjel, Groups of Homotopy Classes. 2nd. revised edition. IV, 36 pages. 1967. DM 4,80 / $ 1.20

Vol. 5: J.-P. Serre, Cohomologie Galoisienne. Troisième édition. VIII, 214 pages. 1965. DM 18,– / $ 4.50

Vol. 6: H. Hermes, Eine Termlogik mit Auswahloperator. IV, 42 Seiten. 1965. DM 5,80 / $ 1.45

Vol. 7: Ph. Tondeur, Introduction to Lie Groups and Transformation Groups. VIII, 176 pages. 1965. DM 13,50 / $ 3.40

Vol. 8: G. Fichera, Linear Elliptic Differential Systems and Eigenvalue Problems. IV, 176 pages. 1965. DM 13.50 / $ 3.40

Vol. 9: P. L. Ivănescu, Pseudo-Boolean Programming and Applications. IV, 50 pages. 1965. DM 4,80 / $ 1.20

Vol. 10: H. Lüneburg, Die Suzukigruppen und ihre Geometrien. VI, 111 Seiten. 1965. DM 8,– / $ 2.00

Vol. 11: J.-P. Serre, Algèbre Locale. Multiplicités. Rédigé par P. Gabriel. Seconde édition. VIII, 192 pages. 1965. DM 12,– / $ 3.00

Vol. 12: A. Dold, Halbexakte Homotopiefunktoren. II, 157 Seiten. 1966. DM 12,– / $ 3.00

Vol. 13: E. Thomas, Seminar on Fiber Spaces. IV, 45 pages. 1966. DM 4,80 / $ 1.20

Vol. 14: H. Werner, Vorlesung über Approximationstheorie. IV, 184 Seiten und 12 Seiten Anhang. 1966. DM 14,– / $ 3.50

Vol. 15: F. Oort, Commutative Group Schemes. VI, 133 pages. 1966. DM 9,80 / $ 2.45

Vol. 16: J. Pfanzagl and W. Pierlo, Compact Systems of Sets. IV, 48 pages. 1966. DM 5,80 / $ 1.45

Vol. 17: C. Müller, Spherical Harmonics. IV, 46 pages. 1966. DM 5,– / $ 1.25

Vol. 18: H.-B. Brinkmann und D. Puppe, Kategorien und Funktoren. XII, 107 Seiten. 1966. DM 8,– / $ 2.00

Vol. 19: G. Stolzenberg, Volumes, Limits and Extensions of Analytic Varieties. IV, 45 pages. 1966. DM 5,40 / $ 1.35

Vol. 20: R. Hartshorne, Residues and Duality. VIII, 423 pages. 1966. DM 20,– / $ 5.00

Vol. 21: Seminar on Complex Multiplication. By A. Borel, S. Chowla, C. S. Herz, K. Iwasawa, J.-P. Serre. IV, 102 pages. 1966. DM 8,– / $ 2.00

Vol. 22: H. Bauer, Harmonische Räume und ihre Potentialtheorie. IV, 175 Seiten. 1966. DM 14,– / $ 3.50

Vol. 23: P. L. Ivănescu and S. Rudeanu, Pseudo-Boolean Methods for Bivalent Programming. 120 pages. 1966. DM 10,– / $ 2.50

Vol. 24: J. Lambek, Completions of Categories. IV, 69 pages. 1966. DM 6,80 / $ 1.70

Vol. 25: R. Narasimhan, Introduction to the Theory of Analytic Spaces. IV, 143 pages. 1966. DM 10,– / $ 2.50

Vol. 26: P.-A. Meyer, Processus de Markov. IV, 190 pages. 1967. DM 15,– / $ 3.75

Vol. 27: H. P. Künzi und S. T. Tan, Lineare Optimierung großer Systeme. VI, 121 Seiten. 1966. DM 12,– / $ 3.00

Vol. 28: P. E. Conner and E. E. Floyd, The Relation of Cobordism to K-Theories. VIII, 112 pages. 1966. DM 9.80 / $ 2.45

Vol. 29: K. Chandrasekharan, Einführung in die Analytische Zahlentheorie. VI, 199 Seiten. 1966. DM 16.80 / $ 4.20

Vol. 30: A. Frölicher and W. Bucher, Calculus in Vector Spaces without Norm. X, 146 pages. 1966. DM 12,– / $ 3.00

Bitte wenden / Continued

MIX
Papier aus verantwortungsvollen Quellen
Paper from responsible sources
FSC® C105338

If you have any concerns about our products,
you can contact us on
**ProductSafety@springernature.com**

In case Publisher is established outside the EU,
the EU authorized representative is:
**Springer Nature Customer Service Center GmbH**
**Europaplatz 3, 69115 Heidelberg, Germany**

Printed by Libri Plureos GmbH
in Hamburg, Germany